Includes DVD

PICTURE YOURSELF as a Magician

Step-by-Step Instruction for the Street, Stage, Parties, Card Table, and More

Wayne Kawamoto

Picture Yourself as a Magician
Wayne Kawamoto

Publisher and General Manager,
Course Technology PTR: Stacy L. Hiquet

Associate Director of Marketing: Sarah Panella

Manager of Editorial Services: Heather Talbot

Marketing Manager: Jordan Casey

Acquisitions Editor: Megan Belanger

Project Editor: Jenny Davidson

Technical Reviewer: Nicholas Carifo

PTR Editorial Services Coordinator: Erin Johnson

Interior Layout: Shawn Morningstar

Cover Designer: Mike Tanamachi

DVD-ROM Producer: Brandon Penticuff

Indexer: Sharon Shock

Proofreader: Sara Gullion

For product information and technology assistance, contact us at

Cengage Learning Customer and Sales Support, 1-800-354-9706

For permission to use material from this text or product, submit all requests online at **cengage.com/permissions**

Further permissions questions can be emailed to **permissionrequest@cengage.com**

Library of Congress Control Number: 2007939374

ISBN-13: 978-1-59863-499-0

ISBN-10: 1-59863-499-2

Course Technology
25 Thomson Place
Boston, MA 02210
USA

Cengage Learning is a leading provider of customized learning solutions with office locations around the globe, including Singapore, the United Kingdom, Australia, Mexico, Brazil, and Japan. Locate your local office at: **international.cengage.com/region**

Cengage Learning products are represented in Canada by Nelson Education, Ltd.

For your lifelong learning solutions, visit **courseptr.com**
Visit our corporate website at **cengage.com**

Printed in the United States of America
1 2 3 4 5 6 7 11 10 09 08

In Memory of Lynn Reinecke Horton, 1956 – 1995

Acknowledgments

Thanks To:

Megan Belanger, for your vision and the opportunity to write this book.

Jenny Davidson, for being my editor. Your timely advice and feedback at every stage of this book was greatly appreciated. Brandon Penticuff, for producing a great-looking DVD. To all of the Cengage Learning staff who lent their artistic, editorial, and marketing expertise to this book.

Nick Carifo, for allowing me to bounce ideas and tap into your in-depth magic expertise, and acting as technical editor for this book. A couple of years back, you allowed me to borrow one of your signature magic routines for a television audition. I didn't get the part. But more important, I knew I had a true friend in magic.

My friends at About.com, in particular Elana Breshgold, who always provides me with timely and useful advice. It's a privilege to write about magic at Magic.About.com.

Lauri Goldenhersh, for letting me know that there was an opening for a writer to cover the world of magic, and all your advice on entertaining.

Kozmo, for inviting me to be a regular commentator for *Reel Magic Quarterly*, the video magic magazine.

Gregg Keizer, for helping me to launch my freelance writing career. You're an awesomely talented writer. Thanks to the many editors who gave me freelance work over the years. I still freelance in the technology and small business areas thanks to Lauren Simonds, Michelle Megna, Forrest Stroud, and Dan Muse.

To my restaurants and bars who provided me with magic gigs. Special thanks to John Petrillo, Judith Walker, and Michael Voeltz.

To all my clients who have welcomed me to their parties and gatherings and into their homes, as well as the many companies, venues, schools, libraries, and cities. It's an honor to entertain at your special events.

To my event planners and agents who have sent me to numerous parties and corporate events. Special thanks to Chuck and Bambi Burnes, Kitty Fisher, Rhonda Bright, Terry Conci, and Hermann and Nancy Leone.

I've had so many friends who have supported my magic, including Tim and Cathy Peron, Ken and Chris Widaman, Deborah Neal, David and Ruth Aldrete, Raymond and Maria Foster, Mark and Kelli Grace Kurtz, Ted and Julia Bacon, Curt and Susan McPherson, Scott and Shelly Wilson, Harry and Ginger Hirakawa, Nick and Cindy Paz, Jim and Susie Elliot, Wayne and Sue Reinecke, and Alan Reinecke.

The Magic Castle and Academy of Magical Arts for providing an incredible venue for watching and researching magic. To the many magicians who provided advice, suggested techniques and tricks, and helped fine-tune routines and moves. There are so many magicians I don't know by name who freely offered me advice at the Magic Castle library. Thanks to Gordon Bean, who provided me with private instruction in my early days, and Andrew Goldenhersh, David Groves, and Jim Skaggs.

Christopher Tindall, for your vision and opportunity to teach magic to your students.

Vincent Lowe and Matteo Lowe, for helping out with the levitation photos in this book.

The many magicians who taught me magic via their books and DVDs, including: Mark Wilson, Jeff McBride, Eugene Burger, John George, Geoff Williams, Jay Sankey, Mike Powers, Daryl, Shaun McCree, Martin Lewis, Dan Tong, John Carney, Thom Peterson, Bob Sheets, Whit Haydn, Bill Abbott, Joe Diamond, Paul Gallagher, Lee Asher, Andrew Normansell, Nathan Gibson, Joshua Jay, Al Schneider, Kenton Knepper, Richard Osterlind, Dan Fleshman, Bill Malone, and Bill Tarr. Your lessons, routines, and techniques are a part of the magic that I entertain with today.

David Copperfield, Doug Henning, Walt Disney, and Ken Burns, for perpetual inspiration.

To my friends Alan Goto and Serge Handschin (did I finally spell your name correctly in this book?) who were around the last time I was writing books. This book is done, let's do lunch. To the Rev. Keith Yamamoto—I always enjoy getting together and talking magic. One of these days, I'm going to watch one of your "fiery" sermons.

My Mom, Chiyo Kawamoto, my uncles and aunts, my grandparents, and my late father, Frank Kawamoto, I was able to attend college and choose careers because of your many sacrifices and hardships.

To Kim Horton and Dana Horton, my nephew and niece. Almost everyone can claim to have a crazy uncle, but you have two. I'm proud of you guys. Sherry Horton, you are family and your business acumen and drive are an inspiration to others.

To my family who put up with all of the magic and my crazy hours. And for this book, there were always pictures to take and new tricks to try out. Thanks for all your patience and assistance.
Love ya, Janet (the best writer in the family), Alysha (the artist), Justin (the singer and actor), and Molly (the equestrian, artist, and actress).

About the Author

Wayne N. Kawamoto is a full-time professional magician who performs at corporate and private events, as well as at restaurants and bars in Southern California. He has performed for the Los Angeles Dodgers and at the Los Angeles County Fair, and entertained at events for Target stores, Nordstrom, Lockheed-Grumman, Countrywide, Century 21, and many more. He is the editorial guide for the Magic.About.com online magazine, where he reviews magic products and teaches magic. His website is www.magicwayne.com.

In prior lives, Wayne has worked as an engineer, project manager, and freelance writer. He has an undergraduate degree in mechanical engineering and an MBA. Prior to becoming a professional magician, he reviewed video games and movies and wrote about small business topics for publications such as *Business Week*, *PC Magazine*, *The San Francisco Chronicle*, *Cinescape*, and more, and published several consumer books on personal computers.

Wayne lives with his family in the community of La Verne, which is near Los Angeles.

Table of Contents

Introduction

I WROTE THIS BOOK BECAUSE when I started out in magic, I couldn't find a book that was right for me. There were the magic books written for little kids with their colorful pictures of top hats and bunnies that explained simple magic tricks, but these books didn't show me how to perform good magic tricks, the kind that my contemporaries might want to watch.

There were the other beginning magic books that were written more for grown-ups and taught pretty much the same simple tricks as the kids' books. And then there were the books that were written for experienced magic enthusiasts and pros that had a language all their own and assumed I already knew and understood the moves and terminology.

I saw the need for an "in-between" book that taught tricks that were baffling and entertaining, yet kept the moves simple so I could get out there and test the magic waters. And I wanted this book to form the foundation of advanced magic that readers would learn later. *Picture Yourself as a Magician* is this book. In addition, the book benefits from its accompanying DVD that demonstrates many of the tricks and techniques in action.

What I've tried to do is select high-impact magic tricks, the ones that professionals often perform, and adapt and simplify them for beginners. The results are tricks that your spectators can enjoy and be mystified by. And you won't have to invest a couple of years in learning the difficult, foundational magic moves and sleight-of-hand, although this book does offer you the necessary starting points.

A Magic Life

I can't tell you how privileged I feel that I now make a living as a full-time professional magician. Magic allows me to be a part of key events in peoples' lives: anniversaries, weddings, retirements, reunions, business milestones, graduations, birthdays, and more. It's an honor to be a part of these important events.

After thousands of hours of performing magic, I have had many memorable moments. There was the time during one of my restaurant gigs when a fellow wanted me to perform a trick and, at the end of the routine, produce an engagement ring so he could pop the question to his girlfriend on the spot. I was excited and told a couple of the servers. The word got out and my magic trick was the focus of the entire restaurant (the stunned young lady enthusiastically said "yes").

I was once hired to perform in a hospital ward for a holiday party for cancer patients and their families. There was so much pain and suffering in that room. I remember as I was setting up for my show that many patients looked weak and exhausted, and I couldn't help noticing all of the medical equipment that surrounded each of them.

But this was a crowd that wanted to be entertained and they responded with surprising vigor. I'll never forget the moment during one of my comedy routines when some patients removed their oxygen masks so they could freely belt out a laugh.

Another memorable moment occurred at a seemingly routine kid's birthday party, the kind that I sometimes perform on Saturday afternoons. The crowd was an average one, but afterwards, an elderly gentleman trudged up to the front where I was packing up my show. He said, "You made me feel like a kid again." That's a compliment I'll never forget.

Great magic provides a means to connect with other people and entertain them. Magic has been very good to me and I hope that I can pass on to you this gift of providing magic and entertainment to others.

Aim high with your magic and you can make just about anything happen.

The Art of Magic

CAN YOU PICTURE YOURSELF as a magician? With this book, you can and will. And you'll become part of a powerful tradition.

Magic is an age-old art that remains relevant in our modern day. Historically, magic was the art of those in power who used tools of deception and trickery to make it appear that they had super-human powers to convince their communities or tribes that they could influence weather, heal the sick, or cause bad things to happen to enemies.

As early as the Middle Ages, magic has had a street tradition that consists of entertainers–who are now known as "buskers"–who performed for crowds and collected money for their work. And for the last 150 years, magicians have been touring the world with large scale and elaborate illusion shows.

With all of the developments in technology and communications, magic has made the leap and remains as popular as ever. Today, Criss Angel performs feats on television and David Copperfield tours with his amazing illusion show. David Blaine continues to come up with feats of endurance that get everyone talking.

Las Vegas, Nevada, and Branson, Missouri are hotbeds of magic entertainment. In Las Vegas, you can regularly see shows by the likes of David Copperfield, Lance Burton, Penn & Teller, Mac King, Hans Klok, Steve Wyrick, Amazing Johnathan, and more. And in Branson, it's Dave and Denise Hamner and Kirby Van Burch. You will also see magicians performing at your local theme park and on cruise ships.

With practice and motivation, you too can learn and perform magic. This book provides everything to get you started.

Here's what's in store.

The Tricks

FROM THE STANDPOINT OF MAGICIANS, there are two types of tricks. There are the tricks that professionals regularly perform. And there are "those tricks"–tricks that are often taught in beginning magic books that no professional or accomplished magician performs.

What sets the tricks in this book apart is that they are not only easy to learn and perform, but they have impact and lots of potential to mystify. To your audience, the tricks may look like those that professional magicians present. Furthermore, the tricks have been selected for their innate entertainment value.

Many of the tricks here are reminiscent of those that have been popularized on television. For example, there's a chapter on levitation that teaches you three ways to levitate yourself, in the vein of famous television magicians. And if you've been following magic on television, you're bound to find some tricks that you have seen. Again, these have been simplified so you can attain success performing them.

Staying with the times and adapting to them, we provide an entire chapter on magic with electronic devices such as iPods, Nintendo DS, and more. We've developed several tricks that will let you perform magic and actually interact with the screens of these devices, which effectively blurs the lines between reality and virtual, electronic worlds.

One more thing, this book itself is magic as there are tricks that you perform with it. There's a card trick that reveals itself in an animation and a mentalism routine that ends in an animation as well–you simply flip through the pages as you would with a cartoon flip book. There are also tricks where spectators will read from this book and follow along.

© istockphoto.com/Kenneth C. Zirkel

As a professional magician, I have included tricks that I would consider performing for my paying clients. Here, you won't find a collection of "those tricks," the dull and castaway tricks found in the other beginning books.

The DVD

THIS BOOK COMES WITH A DVD that allows you to see many of the moves and tricks in action. Many times, when reading about a move or trick, it may seem impossible for it to work or to successfully carry out an illusion and fool spectators. For this reason, the DVD features performances of many of the tricks in this book so you can see them. And once you decide you like a trick, you can refer back to the text to learn the trick or technique step by step. It's the best of both worlds.

Materials

MAGIC CAN BE AN EXPENSIVE HOBBY–there's always a new technique to learn from the latest book or DVD and magic stores often feature cool new tricks that appear to be "must haves." What you'll find in this book are tricks that only require basic household items, such as playing cards, coins, bottle caps, and more. While this may initially sound unimpressive, it's the skills that you apply to these props and the magic that you make with them that turns basic items into magic ones.

There are some props that are explained in this book to build, for example, to perform one of the levitations. But you can purchase the materials from any office or craft store and the materials won't set you back a lot, probably less than ten dollars.

Learning Magic Skills

THIS BOOK PROVIDES AN introduction to sleight-of-hand and features handpicked techniques (moves) that are taught in thorough step-by-step instructions. While there are chapters that allow you to jump in and learn magic that doesn't require any technical skills, we teach fundamental techniques so you can get up to speed and perform baffling card and close-up work.

I've chosen to teach sleight-of-hand moves that are practical and accomplish important purposes. But the moves are also those that can be mastered fairly quickly so you can be successful at applying them. There are also flourishes that can convince spectators that you have skill and will catch their attention.

© istockphoto.com/Eva Serrabassa

Types of Magic

The book features chapters on the different types of magic. The first major distinction is between close-up and stand-up magic. Close-up magic is performed for small groups of people and uses smaller props such as playing cards, coins, finger rings, and more. For most amateur magicians and hobbyists, close-up magic is the type of magic that they perform most.

Stand-up magic is performed from stage or in front of larger groups. Because people have to be able to see what you are doing, this type of magic usually requires larger props and is performed in more formal settings. Stand-up magic often has other props such as tables and you may have a case from which you bring out props. Also, music is commonly used.

Traditionally, street magic referred to shows that were given by entertainers in public locations who later "passed the hat" to collect money. These days, street magic can also describe close-up magic that works in unpredictable environments where you may be surrounded by others and you typically make use of whatever is available, even performing your tricks on the sidewalk if necessary. This is the type of "street magic" that's been popularized by David Blaine and Criss Angel on their television shows.

Because playing cards are such an integral part of magic, particularly the kind that you'll be performing for your friends in informal settings and at parties, there are five chapters dedicated to card magic and tricks, and playing cards are also featured in other sections as well. Chapter 2 teaches you easy card tricks that you can learn and perform right away if you just can't wait to get out there and do some magic.

Chapters 3 and 4 teach you basic card moves and sleight-of-hand. These chapters require an investment of your time but will pay off when you can perform the card tricks explained in Chapter 5. Learning card moves is lots of fun because the moves themselves offer intriguing and deceptive techniques, and as you work at developing your skills, you can clearly see your progress.

Chapter 6 offers a thorough course in trick playing cards. In this chapter, we offer techniques for using trick cards so even those who know about such decks won't suspect that you may be using one.

In Chapter 7, we delve into close-up magic and teach you basic techniques and tricks that use coins, rubber bands, and more. Chapter 8 is devoted to mind reading and making predictions. Chapter 9 offers tricks that you can perform in stand-up situations with larger props.

Street magic is the topic of Chapter 10. Here's where you'll learn edgy magic tricks that can work under almost any condition. And if you're planning to perform magic at an upcoming party, you'll find interactive tricks in Chapter 11 that work well at social gatherings.

If you want to reach for the sky, you'll have fun with Chapter 12, our chapter on levitation. Here you'll learn several techniques for lifting yourself off of the ground. And Chapter 13 is where we teach you tricks with your electronic devices such as iPods and Nintendo DS.

Welcome to the world of magic. *Picture Yourself as a Magician* is the beginning book that I wish had existed when I started in magic. I hope that you find your magic journey to be as exciting and rewarding as I have.

Next Steps

If you want to jump right in and learn some magic tricks, Chapter 2 offers a series of easy magic tricks with cards that require no sleight-of-hand or difficult moves.

© istockphoto.com/René Jansa

2

Easy Tricks

IF YOU WANT TO GET STARTED in magic right away, here are some easy card tricks.

You're probably excited about learning and performing magic and eager to get to it right now. Because of this, in this chapter, we explain a series of ultra-easy card tricks that you can learn and perform without needing any sleight-of-hand (which we'll teach in Chapter 4). And you don't even have to know how to handle cards (which we teach in Chapter 3).

Visualize the Tricks

I RECOMMEND THAT YOU WATCH the performance of each trick on the accompanying DVD before you read its description and secret. For this, please refer to the section titled "Basic Card Tricks" that demonstrates each effect.

Offering Cards for a Spectator to Select

Before you can perform card magic, you need to learn how to offer a deck to spectators so they may choose a card. It's not an actual move or sleight, but a technique that you can just as easily learn on your own.

Hold the cards in the left hand, as shown in Figure 2.1.

Figure 2.1
Holding the deck to begin to offer cards.

With the left thumb, push off cards to the right, which are received by the right fingers and thumb, as shown in Figure 2.2.

Figure 2.2
Offering cards to a spectator.

Continue pushing cards with your left thumb and accepting them with your right fingers and thumb. When you get to the approximate center of the deck, you can stop and allow a spectator to freely choose a card.

Aces High

This easy card trick not only locates the spectator's card, but it produces the four aces. I like this trick because the plot makes it look as if you've failed at finding the selected card. But after you change course and find the spectator's card, there's a second revelation—you've magically located the four aces.

Effect

The spectator chooses a card and returns it to the deck. You snap your fingers and say that something has happened to the selected card. As you spread the deck, a card is found to have turned face up. When asked, the spectator says that it's not his card.

You change course, as if reacting to the situation, and state that the turned-over card is actually an "indicator" card that shows you the location of the spectator's card. You then count that many cards down into the deck.

When the counted-down card is turned over, it is found to be the spectator's card. And when the counted, "in-between" cards are turned over, you show that they are the four aces.

Materials

A deck of playing cards. You'll want to perform this one on a table.

Secret

You've arranged the deck ahead of time to produce the aces and set up the "indicator" card.

Preparation

Beforehand, with no one watching, take out the four aces and a five of any suit. Place the five face up on the bottom of the face-down deck, and then place the four aces face down on the bottom of the deck, after the five. For purposes here, we'll use the five of hearts, as shown in Figure 2.3.

Figure 2.3
Set up the deck with the four aces on the bottom face-down and the five of hearts face up, above them.

Performing the Trick

Set the deck down on the table and allow a spectator to cut the deck and place the top half next to the bottom deck. Ask the spectator to look at the card that he cut to, and return the card to the original top of the deck, as shown in Figure 2.4.

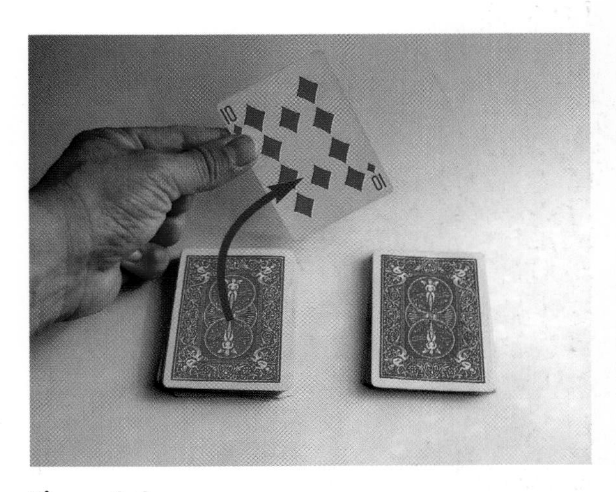

Figure 2.4
Have the spectator cut the deck and then take the card that he cut to and memorize it. Leave the two halves of the deck on the table. (As an example here, the chosen card is the ten of diamonds.)

Place the other pile, formerly the bottom of the deck, on top as shown in Figure 2.5.

Figure 2.5
The spectator returns the card on top of the original top half and then you place the remaining half on top.

> **Reality Check**
>
> At this point, the four aces and face-up five of hearts are on top of the selected card, which resides in the middle of the deck.

Snap your finger and say that something magical has occurred. Spread through the deck and show that the five of hearts has mysteriously turned over and is now face up, as shown in Figure 2.6. When the spectator denies that this is his card, act surprised and say that you forgot that it's simply an "indicator card," which "indicates" that the chosen card resides five cards down.

Figure 2.6
The five of hearts appears face up in the deck.

Count down from the indicator card five cards and prove that the fifth card is the spectator's card, as shown in Figure 2.7.

Figure 2.7
The fifth card down is shown to be the spectator's card.

Pause for a moment and then say that as an added bonus, you've also found the four aces. Turn over the four "in-between" cards to show that they are the aces, as shown in Figure 2.8.

Figure 2.8
Show the "in-between" cards to be the four aces.

The strength of this routine lies in the fact that it looks as if you've failed at finding the selected card. But at the end, you not only come through and find the spectator's card, you also produce the four aces.

If you act a bit flustered as if you've failed to find the card, this should make spectators relax for a second. Then you can hit them with the powerful double revelation at the trick's end.

The Searchers

This is a great effect that predicts two cards. It appears fair because the deck is shuffled by spectators and the dealing is completely legitimate. In the end, despite all of the random factors, you successfully predict the outcome.

Effect

The spectator somehow locates the mates of prediction cards in a thoroughly mixed-up deck. After the deck is shuffled by spectators, you spread the deck and bring out two face-up cards and lay them on the table. You deal cards from the deck and ask a spectator to say "stop," and then insert one of the face-up cards. You continue dealing until a spectator says "stop" again, and insert the second face-up card.

At the end of the trick, you show the spectator that the face-up cards are next to their mates in the deck. For example, if your face-up prediction cards are the seven of hearts and the jack of spades, these cards find their way next to the seven of diamonds and jack of clubs.

Materials

A deck of playing cards. You'll need to perform this one on a table.

Secret

The trick works by itself. Just follow and memorize the instructions. It helps if you read the instructions with a deck of cards in your hands.

Performing the Trick

Hand the cards to a spectator and ask her to mix them in any way. When you get the deck back, turn the cards so that they're facing you and note the top and bottom cards. Quickly browse through the deck and find the mates to these cards. As an example, if the top card is the jack of clubs and the bottom card is the seven of diamonds, you would be looking for the jack of spades and the seven of hearts, as shown in Figure 2.9.

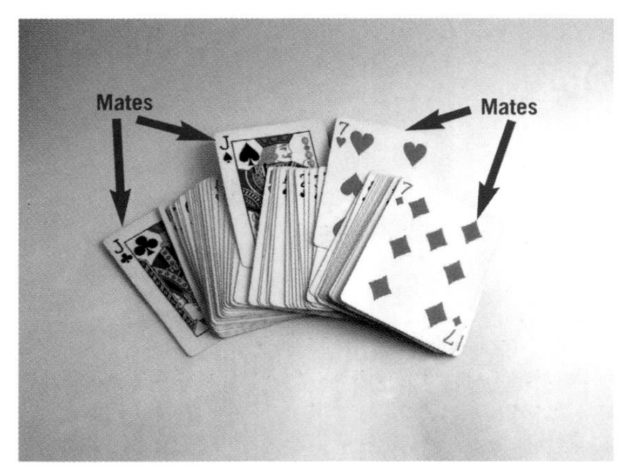

Figure 2.9
Find the mates to the top and bottom cards. In this case, the jack of spades for the jack of clubs on top of the deck and the seven of hearts to match the seven of diamonds on the bottom of the deck.

As you look through the deck, you can tell the spectator that you are looking for prediction cards that are based on a feeling. You can also emphasize that the spectator completely mixed up the cards so there's no way that there is some pre-set order to the cards.

Take the matching "mate" cards and lay them on the table face up. Remember which one is the mate to the bottom card.

It helps if you lay down the mate to the bottom card on top so that it slightly overlaps the mate to the top card. You'll be using the bottom mate first.

Turn the deck face down and begin to deal cards onto the table into a pile. Tell the spectator to say "stop" whenever she wants to. When the spectator says "stop," stop dealing and place the mate to the bottom card (seven of hearts) face up on top of the pile, as shown in Figure 2.10.

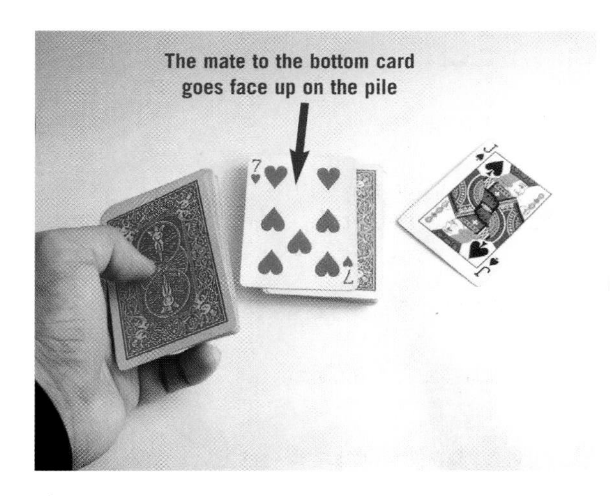

Figure 2.10
When the spectator says "stop," place the mate to the bottom card face up on the pile.

Place the rest of the deck on top of the pile.

Reality Check

At this point, you have secretly placed the face-up seven of hearts next to its mate, the seven of diamonds.

Pick up the entire deck and deal cards once again onto the table and ask the spectator to say "stop" whenever she wishes. When the spectator says "stop," stop dealing and place the mate to the (former) top card (jack of spades) face up on top of the pile and set the rest of the deck on top of the pile. As in the first instance, you have secretly placed the face-up jack of spades next to its mate in the deck.

Reality Check

Because of the way you've dealt the cards, the mates of the original bottom and top cards are now next to each other.

Now it's time for some magic.

Spread through the deck and remove the face-up cards along with the cards directly on top of them, as shown in Figure 2.11. If you like, you can lay the cards down on a table and spread through them to find the face-up cards and the cards directly above them.

Figure 2.11
Look for the face-up cards and remove them as well as the cards directly above them.

Emphasize to the spectator that she shuffled the deck and had a free choice of when to say "stop." But despite this, you can show that the mates to her cards are right next to your predictions, as shown in Figure 2.12.

Figure 2.12
The spectator has mysteriously located the mates to your prediction cards.

This trick is particularly deceptive because everything appears fair and because spectators have no idea what you're about to do, which makes it difficult to backtrack and figure out the secret.

Number Down

Here's a great trick that allows you to tell spectators where in the deck their selected card resides. After a series of cuts, you can tell a spectator how many cards to count down from the top to find his card. When the spectator deals and counts cards, he indeed finds his card. This trick only requires a simple setup that does the work for you.

Effect

A spectator selects a card that is returned to the deck. The spectator cuts the deck and then you cut the deck several times. After all of the cutting, you state that the card is a certain number from the top, say, ten cards. The deck is turned over and cards are dealt. The tenth card is indeed the spectator's selected card.

Materials

A deck of playing cards. You'll need to perform this one on a table.

Secret

A set of cards is in a specific order that will later tell you the location of the chosen card.

Preparation

Take all of the cards of one suit and put them in order, face up, from ace through king on the top of a face-up deck (you're actually placing the cards on the bottom of a face-down deck). For purposes here, the ace equals 1, jack equals 11, queen equals 12, and the king equals 13. As an example, we've used the hearts. You should see the ace on the top of the face-up deck as shown in Figure 2.13. You're ready to go.

Figure 2.13

Take the cards in a single suit and place them, in order, on the top of the deck. Remember that things are reversed as we are looking at a face-up deck. When you turn the deck over, face down, the ace of hearts will be on the bottom.

Performing the Trick

Take out the deck and set it on the table. Ask a spectator to cut the deck and place the top half on the table.

Ask the spectator to look at the top card of the bottom half of the deck and remember it and show it to others, as shown in Figure 2.14. As an example, we'll use the eight of spades.

Figure 2.14
Have the spectator cut the deck and then take the card that he cut to. Leave the two halves of the deck on the table.

The spectator returns the card to the top of the upper half of the deck. Complete the cut, placing the bottom half of the deck on top.

Ask the spectator to cut the cards.

Turn the deck face up. Continue cutting the cards until you see one of the cards in your suit (hearts) on top. Note the numeric value of the heart card that you see (ace = 1, two = 2, three = 3,...ten = 10, jack = 11, queens = 12, and kings = 13). This is your key number and tells you the location of the selected card from the top of the deck.

In our example, we've cut the deck until we revealed a heart, the ten of hearts, as shown in Figure 2.15.

Figure 2.15
Cut the cards until you reveal any heart card. In this example, we've cut to the ten.

In Figure 2.16, we've spread the cards to show the current situation. The selected card is now ten cards from the top. Note that you do not want to show this to spectators. This is for explanation purposes only.

Figure 2.16

This exposed view shows that the selected card is ten down from the top of the face-down deck.

Turn over the deck and pause for a moment and look as if you are concentrating. Announce that the card is so many (insert your number here) cards from the top. Deal the cards one by one and count them. When you get to the designated card, show that it is the spectator's card.

> **Hint**
>
> For dramatic purposes, it's better to count the cards and then hold up the spectator's card with its back to the audience so you don't immediately reveal it. Ask the spectator the name of his card and then slowly turn the card around to show that you've found it.

Four of a Kind—Version 1

A popular effect in card magic is to find the four aces from a mixed deck. In this trick, you won't find the aces, but the spectator will after some cutting and dealing. While the trick as it stands is a decent one, it's far more impressive when you can perform some of the false cuts that are taught in Chapter 4. We'll note the points in the routine where a false cut would work well.

Effect

The spectator cuts the deck into four piles and after dealing some cards, ends up with an ace on the top of each pile.

Materials

A deck of playing cards. You'll need to perform this one on a table.

Secret

This trick works by itself. Just follow and memorize the instructions. It helps if you follow along with a deck of cards in your hands.

Preparation

Remove the four aces ahead of time and place them face down on top of the face-down deck, as shown in Figure 2.17.

Figure 2.17
Secretly place the four aces on top of the deck.

Performing the Trick

Take out the deck and place it onto the table. If you can, perform a false cut here that maintains the order of the deck. Tell the spectators that you are going to cut the deck into four packs.

Demonstrate to the spectators how they will drop a portion from the bottom of the deck (about 25 percent) and then move to the side and remove another quarter of the deck until this is done four times and they've used up the entire deck. Figure 2.18 shows how to cut the deck into piles.

> **Reality Check**
>
> The four aces sit on top of the fourth pile that you cut—the remainder of the pack after cutting the deck and dropping a portion of it on the table three times.

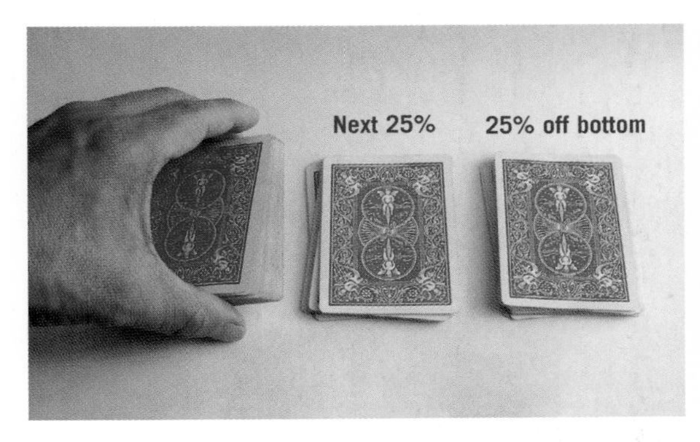

Figure 2.18
Cutting the deck into piles for the trick.

You'll reassemble the deck in an offbeat way that looks as if it is going back together in a random fashion but is secretly putting the top quarter of the deck back to the top.

Using the number system as shown in Figure 2.19, stack 1 (the stack with the aces) on 3, then 2 on 4. You'll now have two piles. And then place the combination of 1 and 3 on top of 2 and 4. The unexpected stacking sequence looks random, but you'll end up with the four aces on top of the deck once again.

Figure 2.19
The restacking sequence makes it look as if you've mixed the cards.

From here, you won't touch the cards.

Ask the spectator to cut the deck in the same fashion as you did, dropping a quarter of the deck at a time until there are four piles. Again, and unknown to the spectators, the four aces will reside on top of the fourth pile, the remainder of the deck at the end.

> **Note**
>
> Sometimes spectators stray from your instructions. Just follow where the aces lie and adjust accordingly. Just be sure that the last pile the spectator deals from has the aces.

Tell the spectator to pick up a non-ace pile and deal three cards from his pile to the table where the pile resided, and then deal a single card on top of each of the other piles, as shown in Figure 2.20.

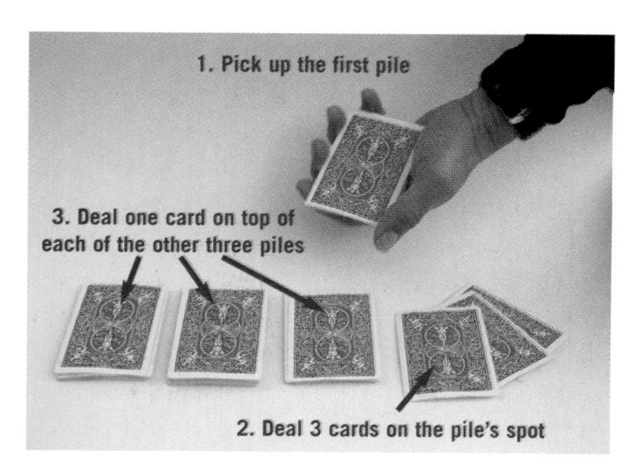

Figure 2.20

Have the spectator grab the first pile, deal three cards into the space where the pile formerly resided, and then deal one card on top of each of the other piles.

Tell the spectator to take the remaining cards and rest them in their original location, on top of the three dealt cards.

Tell the spectator to take a second non-ace pile and do the same thing—deal three cards to the table in its original, now empty, location, and then deal a single card on top of each of the other piles.

Tell the spectator to take a third non-ace pile and do the same thing—deal three cards to the table and then deal a single card on top of each of the other piles.

> **Reality Check**
>
> The aces lie three cards down from the top in their pile.

Tell the spectator to take the final pile, the one with the aces, and do the same thing—deal three cards to the table and then deal a single card on top of each of the other piles.

> **Reality Check**
>
> There's an ace on top of each pile.

At this point, emphasize how the spectators did all of the cutting and dealing and that there is no way that you could have controlled the process.

Ask the spectator to turn over the top card of each pile to show that there is an ace on top of each, as in Figure 2.21.

Figure 2.21
The spectator has cut and dealt cards so four aces are on top of each of four piles.

Four of a Kind—Version 2

While the results of this effect are similar to the previous one, the method is quite different. Between the two, I personally prefer this one because the steps feel more natural and not as convoluted as those in the first version. You can choose to produce any four of a kind. For purposes here, we'll produce the four jacks.

Effect

After some dealing, the spectator ends up with four piles, each of which has a jack on top.

Materials

A deck of playing cards. You'll need to perform this one on a table.

Secret

The trick works by itself. Just follow and memorize the instructions. It helps if you follow along with a deck of cards in your hands.

Preparation

Remove the four jacks ahead of time and place them face down on top of the deck, as shown in Figure 2.22.

Figure 2.22
Place the four jacks on top of the deck.

Performing the Trick

Take out the deck and place it onto the table. If you can, perform a false cut here.

Give the deck to the spectator and ask her to deal the cards into two piles on the table, alternating back and forth between the two piles. After about five cards are dealt into each pile, ask her to stop at any time that she wishes. When the spectator stops, ask her to put down the deck. You'll have the setup shown in Figure 2.23.

Figure 2.23
The spectator has dealt two piles onto the table.

> **Reality Check**
>
> The jacks are on the bottom of the two piles—two jacks at the bottom of each pile.

Pick up one of the piles and ask the spectator to pick up the remaining pile. Begin to deal cards from your deck into two piles on the table and ask your spectator to do the same.

When both of you have dealt all of the cards, you'll end up with four piles of cards on the table.

> **Reality Check**
>
> The dealing reverses the cards so you now have jacks on top of each pile.

Ask the spectator to turn over the cards on the top of each pile. She'll find that they are all jacks, as shown in Figure 2.24.

Figure 2.24
The spectator has dealt cards and ended up with jacks on top of each pile.

The Prediction

The Effect

You bring out an envelope. The spectator mixes up the deck and deals cards onto the table. When the spectator opens the envelope, he finds the name of a card. And when he turns over the top card of the dealt pile, he finds the named card.

Materials

A deck of cards and an envelope with a piece of paper inside. You'll also need a table to perform this trick.

Secret

You introduce the top card after the cards have been dealt with a sneaky move.

Preparation

Remove a card from the deck that you want to use as the prediction card. Write the name of this card on the piece of paper and fold it and place it inside of the envelope. If you like, you can seal the envelope. Rest the envelope on the table with the prediction card hidden under it, as shown in Figure 2.25.

Figure 2.25

Rest the envelope on the table with the prediction card hidden under it.

Performing the Trick

Make sure that the prediction card is face down and that you can easily pick up the prediction card and envelope together when it's time to use the envelope. Some performers like to extend the envelope and card over the side of the table a bit so the two objects are easier to grab and hold together.

With your envelope and card in place, take out the deck of cards, hand it to a spectator and ask them to mix the cards.

After the spectator has finished mixing the cards, ask him to deal cards onto the table. After he has dealt eight cards, tell him to stop at any time. This is shown in Figure 2.26.

Figure 2.26

The spectator deals cards until he feels like stopping.

When the spectator stops dealing, grab your envelope with the prediction card and carry it over to the deck. Rest the envelope on the deck and with it, secretly rest the card underneath it on the deck.

Be careful that you don't tilt the envelope and allow spectators to see the underside with the card, as in Figure 2.27.

Figure 2.27
Lift the envelope and the card and place it on the pile of cards. Note that this is an exposed view. You do not want the spectator to see you placing the hidden card onto the deck.

> **Reality Check**
>
> Your predicted card that you've written on the paper now lies on top of the deck. You've secretly introduced it after the spectator has dealt the cards.

Ask the spectator to pick up the envelope and open it and read the note that's inside. After the note is read, emphasize to the spectator that you haven't been near the cards.

Ask the spectator to turn over the top card of the deck. When the spectator turns over the top card, it's found to be the same one that you had written down on the note, as shown in Figure 2.28.

Figure 2.28
The written prediction matches the card that the spectator selected.

The Reversed Card

Magicians love card effects where a chosen card reverses itself in the deck. This trick is one that almost every magician learns early on. And the concept is the foundation for a host of tricks with face-up and face-down cards.

The Effect

A spectator selects a card and returns it to the deck. The magician takes the deck and places it momentarily behind his back. He brings back out the deck and when he spreads through it, shows that one card is facing the wrong way. The reversed card is the spectator's selected card.

Materials
A deck of cards.

Secret
By reversing a card, you make the entire deck appear to be facing the opposite direction.

Preparation
Invert the bottom card of the deck before you perform the trick—the bottom card of the deck faces the other way—face up in a face-down deck, as shown in Figure 2.29.

If you like, you can set the deck up and then place it into a card box. Just remember which side is the real top of the deck.

Figure 2.29
Secretly reverse the bottom card of the deck. Note that this is an exposed view for demonstration purposes. You would not want spectators to see this.

Hint

When performing card tricks where I have to remember which is the "right" orientation for the deck, I position the deck so its back is on the same side of the card case that displays the back design, as shown in Figure 2.30.

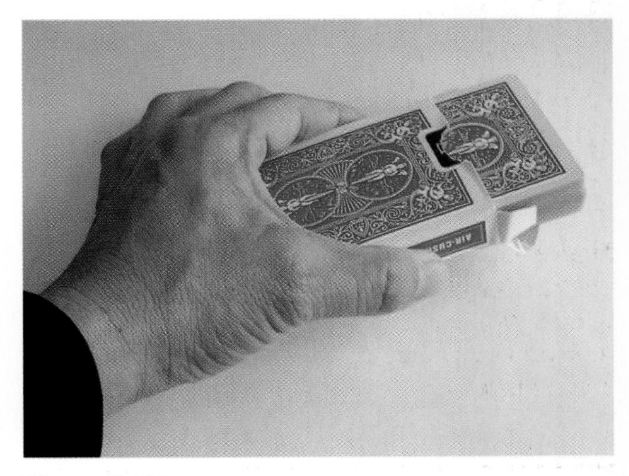

Figure 2.30
Match the correct orientation of the deck with the back design on the card case.

Performing the Trick
Take out the deck of cards, spread the deck and allow the spectator to freely select a card. Be careful not to allow the wrong-facing bottom card to show.

As the spectator looks at the card and shows it to others, casually turn the deck in your hand so that the bottom card is now on top. The deck will look normal except that it's upside down, with the bottom (now top) card facing the wrong way, as shown in Figure 2.31. It's important that you keep the individual cards in the deck together and don't allow them to separate and reveal that the deck, with the exception of the top card, is facing the opposite direction.

Figure 2.31
The bottom card that is reversed is now the top card. It makes the deck look normal except all of the other cards are reversed. Note that the deck has been spread a bit for demonstration purposes.

Ask the spectator to return the card to the deck. Since you can't allow the cards to be separated to receive the selected card, it's a good idea to grasp the card as it's being returned and guide it into the deck. You'll have to keep the deck closed so spectators don't see that the deck is upside down and use more force than normal to return the card into the deck. This is shown in Figure 2.32.

Figure 2.32
Take extra care when returning the spectator's card to not allow the deck to separate and expose the fact that most of the deck is facing the opposite direction.

Take the deck and place it behind your back. You can explain to spectators that you are now performing "magic" without looking. Or you can say something along the lines that your fingers are very sensitive and behind your back you're attempting to find the card.

When the deck is behind your back, turn over the bottom card so it's facing the same direction as the rest of the deck.

Reality Check

The spectator's chosen card, because it was inserted the wrong way, is now the only card that is reversed in the deck.

The trick is done. Bring the deck back out.

Hold the deck face up and begin to spread through the deck. You'll find a card that is oriented the other way, face down. Show that there is only one reversed card in the entire deck.

Ask the spectator what his card was and show that his card was reversed in the deck, as shown in Figure 2.33.

Figure 2.33
The spectator's card has mysteriously reversed itself in the deck.

Next Steps

In Chapter 3, you'll learn the basics of dealing and shuffling cards and master a flashy one-handed cut.

After establishing a foundation in basic card handling, you can move onto Chapter 4, "Basic Card Sleight-of-Hand," where you'll learn variations that will allow you to control the location of cards, appear to cut and mix the deck when you are not changing its order, and more.

Basic Card Handling

YOU HAVE TO BE ABLE TO STAND before you can walk. And in much the same way, you need to know how to handle playing cards, which will form the foundation for your card magic.

In this chapter, you'll learn the basics of dealing and shuffling cards and learn a flashy one-handed cut.

While learning basics often conjures up images of boring and repetitive exercises that are designed to get your fingers and/or body in shape to prepare for what's next, you'll find that the card handling skills here are techniques that you can immediately put to use, whether you're playing cards with friends or performing some of the simple, self-working card tricks that you may have learned in Chapter 2.

Later on, you'll be building on your card skills. For example, you need to know how to perform a real shuffle that mixes the cards before you can execute a fake one that doesn't. Begin with the techniques here and work on mastering them until you can execute them smoothly and flawlessly.

Visualize the Card Handling

YOU MAY KNOW SOME of the card handling techniques already. But before you move onto the nuts and bolts of learning and executing the various shuffles, cuts, and other techniques in this chapter, you may find it helpful to view the techniques in action on the accompanying DVD. For this, please refer to the "Basic Card Handling" video that demonstrates each technique. After watching the video, you may discover that there are techniques you want to learn first, and you can move directly to the appropriate section.

What Type of Playing Cards Should You Use?

While it's perfectly fine to use whatever playing cards you happen to have around, you'll probably want to settle on a particular brand of playing cards that will respond in a consistent way. When you're just starting out in magic, it's a good idea to try out different playing cards and determine which ones you like the best.

Evaluate how the cards feel in your hands and how they work with your moves. Are the cards too big? Too small? Are they too slippery or do they stick together? In general, plastic coated cards are not good for magic because they tend to stick together. This isn't as important when you're performing the easier tricks in Chapter 1. But when you get into sleight-of-hand, the stickiness can impede some moves.

Another consideration: cheap cardboard playing cards will not work well for magic as they tend to tear after only limited use.

Personally, my choice of brand was driven by feel and economics. I like the finish on Bicycle cards and initially preferred the smaller "bridge" size cards because I found them easier to hold and work with. However, my local warehouse store sold poker-sized Bicycle cards in bulk for as little as a $1.25 each—less than half the price of Bicycle cards available elsewhere.

As a result, I have always used these cards and grown accustomed to working with them. I happen to like the feel of Bicycle playing cards, and when I work with other brands, even high quality cards such as the U.S. Playing Card Company's "Bee" cards, I find that there are moves that are hard to perform because of the different finish.

Another advantage with using common Bicycle cards is that everyone has seen them and spectators are less likely to suspect them to be trick cards. As you'll learn in later chapters, Bicycle cards can be purchased in trick versions, which will work to your advantage.

Dealing Cards

THE ABILITY TO DEAL CARDS is fundamental to performing magic as well as playing card games. Chances are, you already know how to deal cards, but here are a few thoughts with regard to card magic.

Most people deal the cards using their dominant hand and hold the pack in the other. If you look at the picture in Figure 3.1, you can see how the deck is held—actually it's more like cradled—in the hand. Because I'm right-handed, I hold the deck in my left hand. If you're left-handed, reverse the instructions.

Figure 3.1
Hand position for dealing cards.

Notice how the first finger rests on the top edge of the deck, and the thumb along with the first and second fingers cradle the long sides. In later chapters, you'll see the important role that the little finger plays.

To deal, the left thumb pushes off the top card—in this case, to the right—and the top card is picked up by the thumb and first and second fingers of the other hand, which deal the card to the table. This is shown in Figures 3.2 and 3.3.

Figure 3.2
The left thumb pushes off the top card.

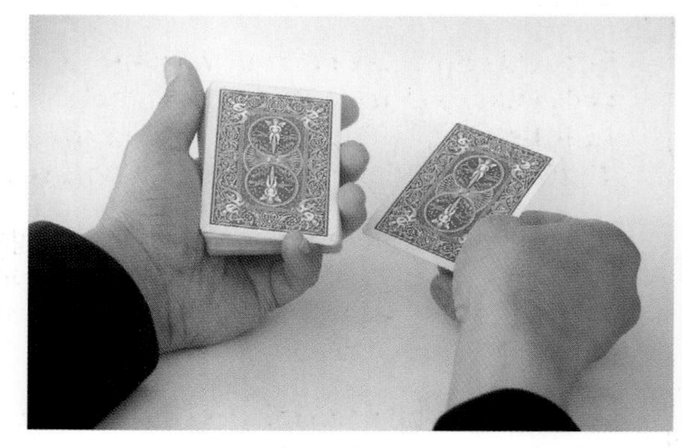

Figure 3.3
Dealing cards.

The Overhand Shuffle

THIS WELL-KNOWN SHUFFLE changes the order of the deck by mixing cards in groups. It's much like executing a series of small cuts to the deck. It's a standard shuffle among card players and, as a result, is a particularly good one to use in your magic.

If you're right-handed, you'll hold the cards in your right hand, with the second and third fingers holding the top (short) edge and the thumb holding the bottom edge, as shown in Figure 3.4. Your left hand lies waiting, ready to receive the portions of the deck that will be dropped on their sides.

Figure 3.4
Getting ready to perform an overhand shuffle.

As shown in Figures 3.5 and 3.6, the right hand moves up and down and on the downward strokes, the fingers and thumb drop off a group of cards from the top of the deck and leave them in the left hand. The right hand carries the deck up and then drops down and leaves more cards from the top of the deck onto the pack that's forming in the left hand. The left thumb moves out of the way to allow the new cards to fall through and then rests against the deck to retain the cards.

Figure 3.5
The right hand delivers groups of cards to the left hand.

Figure 3.6
Continue to deliver groups of cards to the left hand until you've used up the pack.

When you get to the bottom of the pack, you simply throw the remaining cards onto the pack in the left hand to complete the shuffle.

As you improve, you may notice that your left hand will also move back and forth to meet and pull away from the right hand. This is fine.

The number of cards that you drop on each stroke is up to you. You can finish an overhand shuffle in as little as three or four "drops." Some people like to perform as many as 15 or more drops.

Try to aim for a smooth and quick action. When done right, an overhand shuffle can be performed in a few seconds.

The Hindu Shuffle

THE HINDU SHUFFLE is essentially an overhand shuffle that's performed with the cards held on the long ends. If you've never seen this shuffle, it's because it's only popular among magicians. By turning an overhand shuffle on its side, the Hindu, as you'll discover in later chapters, allows for various magical manipulations.

I've never seen anyone use a Hindu shuffle in a game of cards. Despite the fact that most spectators have never before seen a Hindu shuffle, I often use one and none of my spectators have ever questioned it.

As shown in Figure 3.7, the deck is held face down with a long side towards you. With your right hand, grip the front-most long edge of the deck with the thumb and grip the back edge with the rest of the fingers, mostly the first and second fingers. The left hand lies open underneath and to the left of the deck, ready to receive the cards.

The description given here is how I happen to hold the cards; however, there are variations. Some magicians like to hold the top part of the deck with the second and third fingers on the back side and rest the first finger on top to better manage the deck. Feel free to experiment with the grip and discover what works best for you.

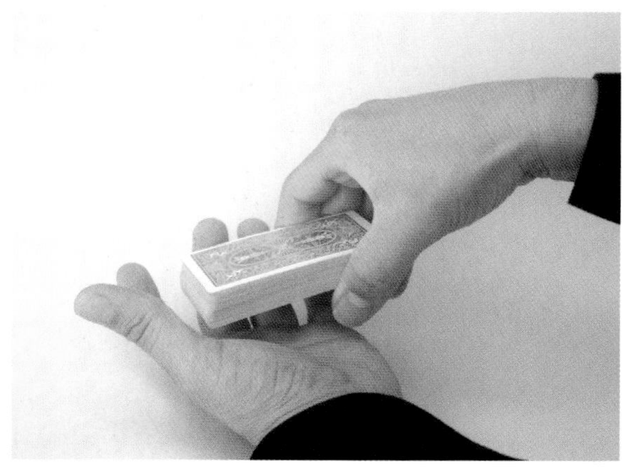

Figure 3.7
Getting ready to perform a Hindu shuffle.

The right hand moves the pack into the left hand, and the thumb and first and second fingers of the left hand grip a small group of cards from the top of the deck along the long edges. The right hand pulls away and the thumb and fingers of the left hand momentarily hold the group of cards and then allow them to fall down into the hand. The process is shown in Figures 3.8 and 3.9.

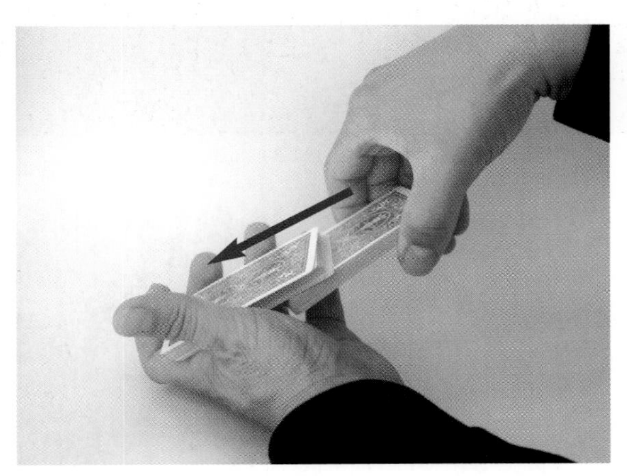

Figure 3.8
The right hand delivers groups of cards to the left hand.

This is repeated until you've run through the entire deck.

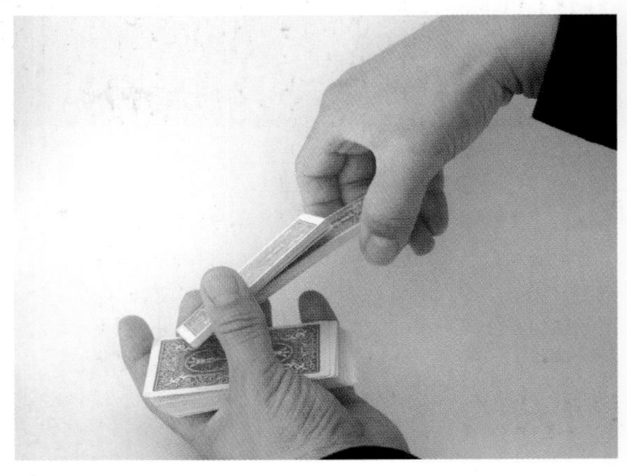

Figure 3.9
Continue to deliver groups of cards to the left hand until you've exhausted the pack.

The final packet that remains in the right hand is simply thrown on top of the pack in the left.

The Riffle Shuffle

THE TERM "RIFFLE" IS ONE that magicians use and is not normally used by non-magicians. *Riffle* refers to the process of thumbing or "riffling" through the cards—much like using your thumb to run or flip through the pages of a book.

This shuffle is better known among the general public as a *dovetail shuffle*. The deck is separated in halves and the cards are run off of the thumbs of each hand ("riffling") so that the cards interlace and mix. This well-known shuffle has several variations.

As the name implies, the *table* version of the riffle shuffle is performed on a table and is often used by poker players and gamblers. In this version, the interlace, the process of weaving the cards together, is often performed at an angle—the two halves are usually around 120 degrees relative to each other. The shuffle performs a thorough mix, and it is performed low on the table to prevent anyone, especially the dealer, from spotting cards as they go by and later recalling their location. Once the cards are mixed, the two packs are simply shoved back together.

In another popular version, the deck is divided into two halves and interlaced straight on. After the cards are riffled and mixed, you can either shove the cards together as you would in a table riffle shuffle, or perform a technique known as a "bridge" to coalesce and blend the cards together in a flashy manner.

The Table Riffle Shuffle

Lay the deck on the table face down with a long side toward you. Using your right hand, grab the top half of the deck with your right thumb and first and second fingers, as shown in Figure 3.10. If you like, you can maintain a grip on the lower portion of the deck using your left hand.

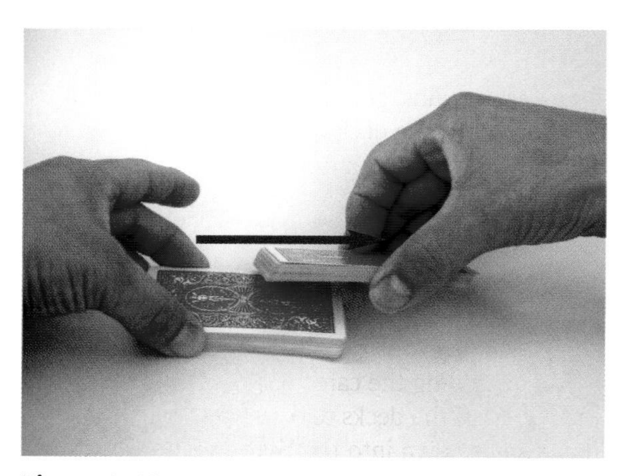

Figure 3.10
Grip the top half of the deck with your right hand.

Carry the top half to the right and lay it down next to the bottom half of the deck. The cards will be in the position shown in Figure 3.11.

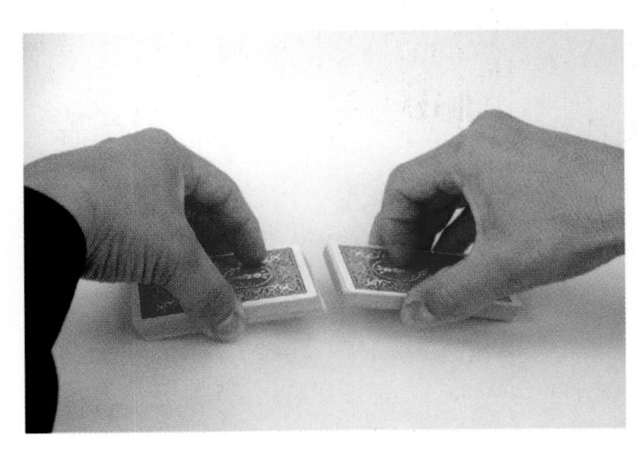

Figure 3.11
Positioning the halves so they may be mixed together.

Using both hands, grab the outer-most long edges of each half of the deck with your third, fourth, and little fingers, and grab the inner-most long edges with your thumbs. Curl your first finger so the tip presses against the back of the cards, as shown in Figure 3.12. Your hands are performing the same actions on their own respective halves.

If you're holding the cards correctly, you can cause the halves of the decks to bow by pressing your first fingers down into the back of each half deck, which is shown in Figure 3.12. This is important to allow the cards to riffle off of the thumbs.

Figure 3.12
Bowing the cards so they may be "riffled" off of the thumbs.

Bring the halves together and lift up the inner corners of the halves using your thumbs. Riffle the corners of the cards in each half of the deck so that they fall and interlace, as shown in Figure 3.13.

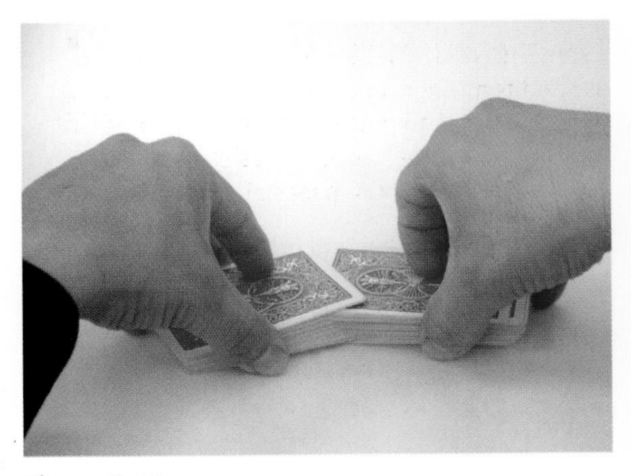

Figure 3.13
Let the cards fall together.

Grip each half of the deck with your hands and let your little fingers rest against the short sides of the deck, opposite the point where the interlace occurred. Push the cards together, as shown in Figure 3.14.

Figure 3.14
Pushing the cards together.

The Straight Riffle Shuffle

There are two ways to get into this riffle shuffle. I'll explain both, but I favor the second method because it's flashier and orients the cards in a manner that is useful for some card tricks. The second method is slightly harder to learn, but well worth the effort.

Separating the Deck—Method 1

Hold the deck with a short end toward you. Using both hands, your thumbs grip the nearest short edge while your second fingers grip the far edge, as shown in Figure 3.15.

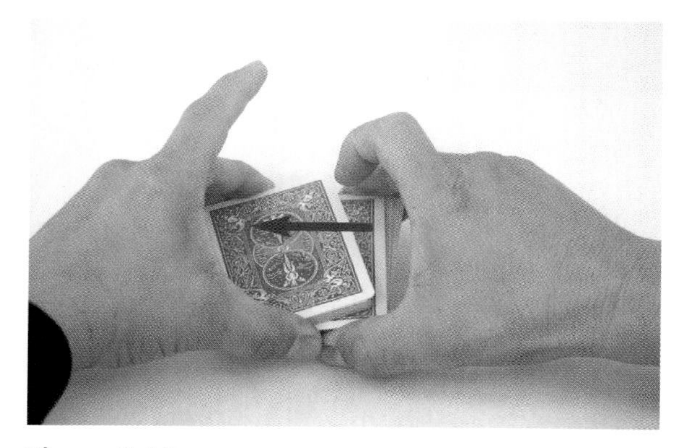

Figure 3.15
Holding the deck before dividing it into two halves.

The second finger of the right hand grips the lower half of the deck on the short edge that's furthest from you, while the second finger of the left hand grips the upper half of the deck on the same short edge. This is better seen from the front, as in Figure 3.16.

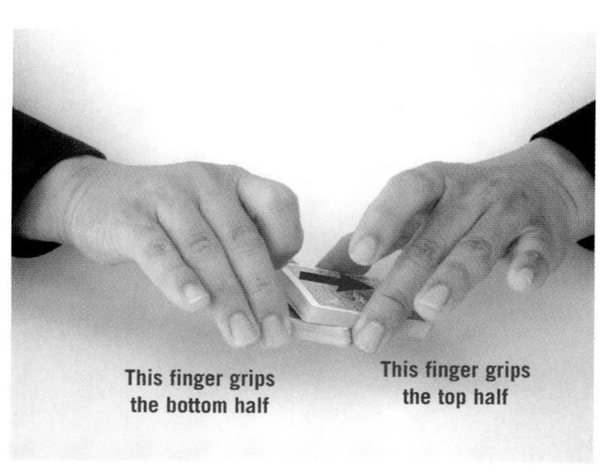

Figure 3.16
A front view of the "split."

By allowing the second fingers to grip different portions of the deck, you can break the deck in half: The upper half is carried away by the left hand and the lower half is carried away by the right hand. This is shown in Figure 3.17.

Figure 3.17
The two halves are turned so that the short edges face each other.

The halves of the deck are positioned so the thumbs are next to each other, as in Figure 3.18. You're ready to perform the interlace.

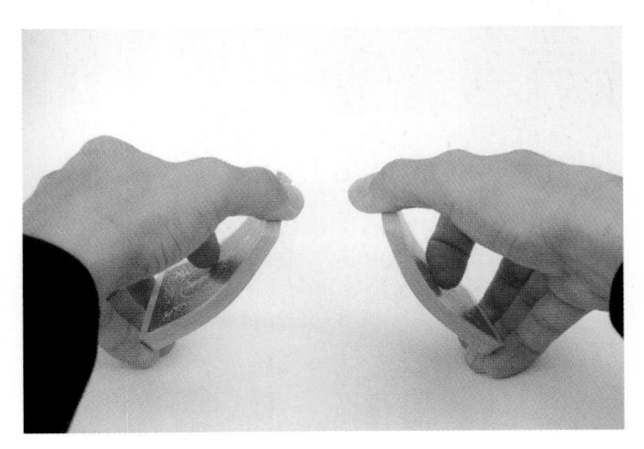

Figure 3.18
The two halves are bowed and you're ready to "riffle" the cards.

Separating the Deck—Method 2

Position the deck vertically in the right hand and hold it as in Figure 3.19. The breakdown is as follows: The thumb holds the top short edge, the second and third fingers hold the bottom short edge, and the first finger is curled against the back of the deck. If you're holding the deck correctly, you can cause the deck to bow by pressing your first finger down into the back of the deck.

Your left hand lies open with the fingers spread out, ready to catch the cards.

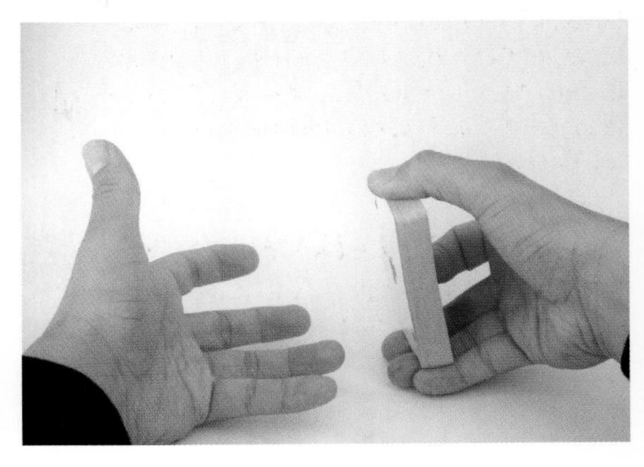

Figure 3.19
Ready to perform a riffle to separate the deck into two halves.

The right thumb riffles the lower half of the deck into the left hand. The falling pack of cards should hit the joint at the base of the left hand's second and third fingers, as in Figure 3.20.

Figure 3.20
The left hand catches half of the deck that's been riffled down.

Once you've riffled the lower half of the deck into the left hand, the second and third fingers of the right hand touch the rightmost edge of the lower half and lever it up, as in Figure 3.21.

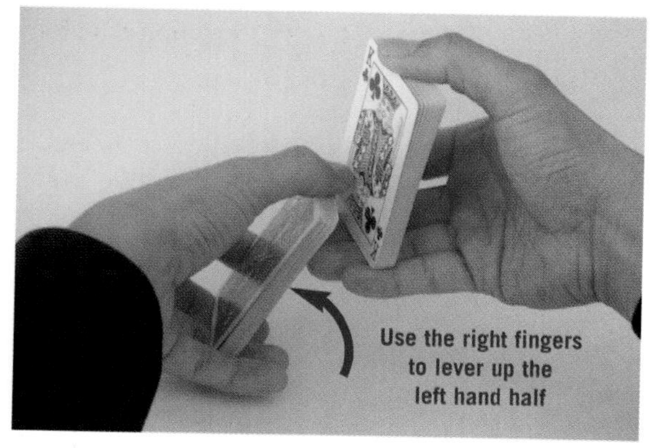

Figure 3.21
Levering the lower half of the deck up.

Grip the lower half with the left hand. You will be holding half of each deck in an identical manner in each hand with the second and third fingers on the bottom short edges, the thumbs on the top short edges, and the first fingers curled into the back of the decks. Your hands will be in the position shown in Figure 3.22. You're ready to perform the interlace.

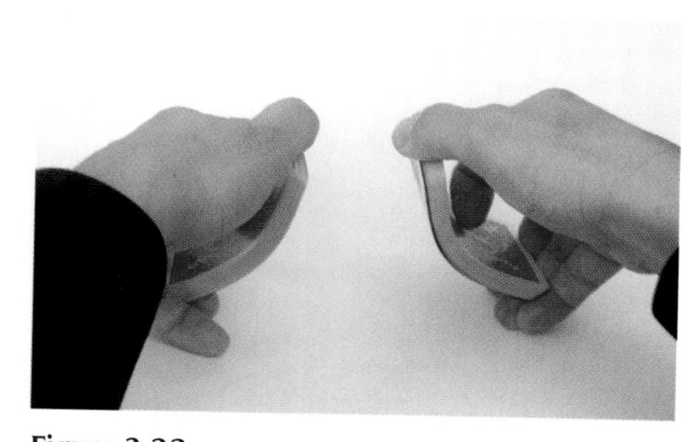

Figure 3.22
Ready to interlace the cards.

The Interlace

With the halves of the deck bowed in each hand, position the hands so when the cards riffle off of the thumbs, their innermost short sides will overlap by a quarter to a half-inch. Bow the cards by pressing down with the first finger of each hand and allow the cards to fall off of the thumb and interlace.

After the interlace, grip the halves of the deck with the fingers of each hand on the bottom of each half. The outside short edges are resting against the joints at the bases of the fingers.

If you would like to "bridge" the cards, continue onto the next section. At this point, you can shove the cards together to complete the shuffle.

The Bridge

After the interlace, grip the halves of the deck with the fingers of each hand on the bottom of each half. The thumbs are on top of each half. The cards should overlap approximately half an inch, as in Figure 3.23.

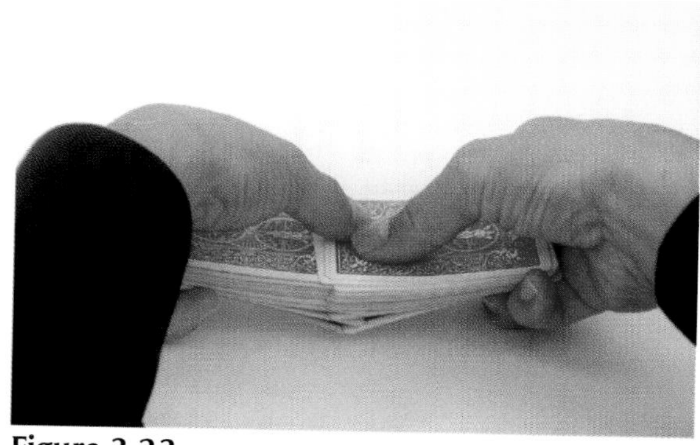

Figure 3.23
Ready to bridge the cards.

Turn the fingers up to flex and bow the cards, as in Figure 3.24.

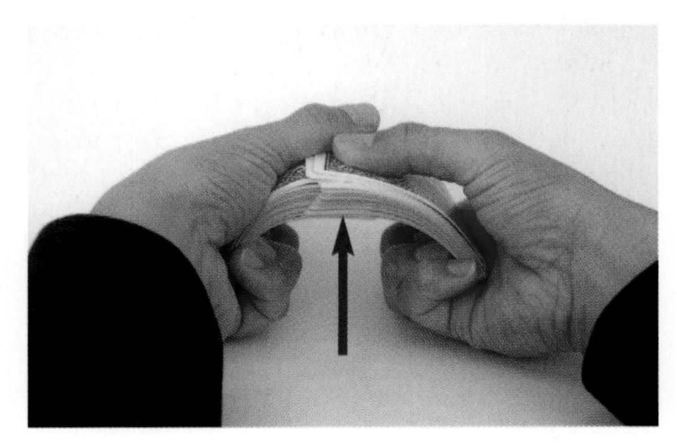

Figure 3.24
Bridging the cards.

There's a definite knack to this that will take practice. Slowly straighten out your fingers and release the pressure that the base of the fingers is exerting on the bowed halves. This allows the halves to fall and coalesce into a single pack. Figure 3.25 shows the bridge "falling" and the cards merging together. You have completed the shuffle.

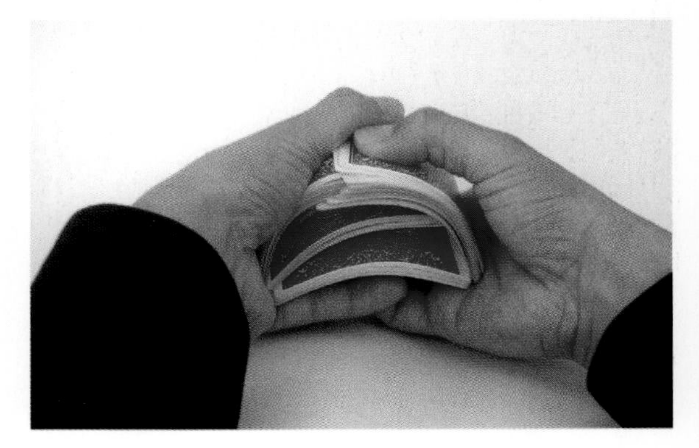

Figure 3.25
The cards fall together from the bridge.

The One-Hand Cut

IF YOU'VE MASTERED THE SHUFFLES taught in this chapter, you'll be considered a competent card handler. However, if you want to stand out from the crowd and show off some digital dexterity, you'll want to learn a flashy one-handed cut.

This particular one-hand cut is known among magicians as the *Charlier*, a French card shark and street performer who popularized its use in the late 1800s. It is not known if Charlier actually invented the cut. The Charlier divides a deck of cards into two halves and makes them change places.

Hold the deck with the thumb against a long edge of the deck and the fingers against the other long edge, as in Figure 3.26.

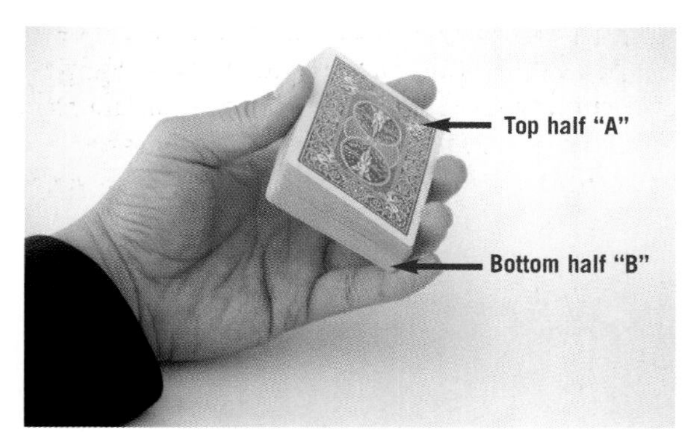

Figure 3.26
Gripping the deck for the one-hand cut.

By releasing pressure on the bottom part of your thumb that's in contact with the deck, drop the bottom half of the deck into your palm, as shown in Figure 3.27.

Figure 3.27
Release the bottom half of the deck.

Release the grip of your second finger and use it to reach under the top half of the deck and contact the lower half of the deck. Push the lower half up towards and against the thumb and allow it to pivot on the palm, as shown in Figure 3.28. Some magicians prefer to release the grip of the first finger and use it to pivot up the half of the deck. Feel free to experiment to identify the technique that works best for you.

Figure 3.28
Use the second finger to push up the lower half.

Once the lower half clears, allow the top half to fall into the palm, as in Figure 3.29.

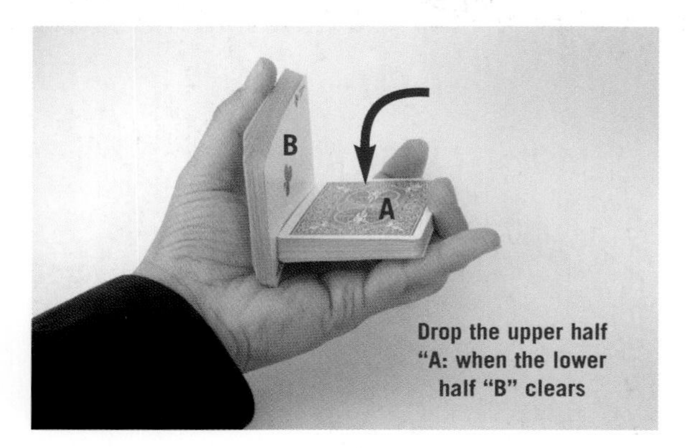

Figure 3.29
The lower half clears and allows the top half to fall.

Push the thumb forward to allow the half that is resting against it to fall onto the other half, thus completing the one-hand cut (see Figure 3.30).

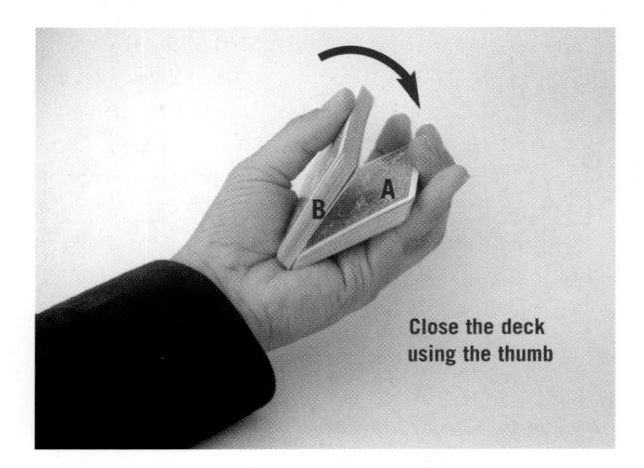

Figure 3.30
Finishing the one-hand cut.

Easier Practice

When learning the one-handed cut, it helps to secure the halves of the deck with rubber bands so you can learn and practice the motions without dropping individual cards and having to pick them up. When you've got the movements down, remove the bands and start with a portion of the deck, say half the deck, which is easier. As you improve, add cards until you're performing the cut with a full deck.

Next Steps

You can use your new card handling skills to dress up the self-working card tricks found in Chapter 2.

After establishing a foundation in basic card handling, you can move onto Chapter 4, "Basic Sleight-of-Hand with Cards," where you'll learn variations on these moves to control the location of cards and other useful techniques that will apply to your card magic.

Basic Sleight-of-Hand with Cards

NOW THAT YOU HAVE MASTERED CARD-HANDLING techniques, it's time to learn some basic sleight-of-hand with cards that will form the foundation of your card magic.

The term "sleight-of-hand" has an almost mystical connotation to the general public—it's virtually a universal theory after "sleeves" and "mirrors" that's used to explain a magic trick. Few lay people can actually define what "sleight-of-hand" actually is. It's a catch-all phrase, even, to a certain degree, among magicians.

"Sleight-of-hand" encompasses the "moves" and skill with our hands that allow us to perform magic. It's the stuff that spectators don't see, the behind the scenes mechanics that makes magic happen. While you learned some good self-working cards tricks back in Chapter 2, here you'll be learning skills that will let you perform even more baffling magic. And we'll be building on the skills that you learned in Chapter 3.

As I like to tell my audiences, if some of the flourishes and other moves that I perform look difficult, there are moves that they don't see that are equally or more difficult to learn and execute. It's true that there are magic moves that can take months or even years to master.

Sleight Difference

WHEN PERFORMING CARD TRICKS, almost all professionals rely on some or a lot of sleight-of-hand. The advantage of sleight-of-hand is that you can perform magic with a deck of normal playing cards. Spectators may suspect that you are using trick cards when you're using real ones, and when you rely on sleight-of-hand as your secret, you can freely hand cards out so spectators may examine them. On the other hand, when you're using trick cards, you're always trying to hide this fact.

Another advantage with sleight-of-hand is that you can perform several tricks using a single deck of playing cards. If you rely on a trick deck that only does one thing, you pretty much only have one trick. As you'll see in Chapter 5, there are lots of tricks that you can perform with just the techniques that you'll learn here.

To perform sleight-of-hand, your hands don't have to necessarily move quickly, although there are advanced moves where speed is crucial. The moves that I have chosen to teach you here are those that are relatively easy to learn and perform, and best yet, will give you the tools to successfully perform sleight-of-hand-based card magic.

It's a foundation that will allow you to get out there and perform. And if you're so inclined, provide a base from which you can pursue more difficult moves to suit your personality and performance style.

So take your time. Work on mastering the techniques here until you can execute them smoothly and flawlessly. Most of the moves are built on the handling skills that you learned in Chapter 3.

Visualize the Card Handling

Before you start your journey into sleight-of-hand, you may find it helpful to view the techniques in action on the accompanying DVD.

For this, please refer to the "Basic Card Sleight-of-Hand" video that demonstrates each technique. After watching the video, you may discover that there are techniques that you want to learn first and can move directly to the appropriate section.

Card Controls

IN CARD MAGIC, we control the location of a card by either taking a card to a known location such as the top or bottom of the deck or putting something next to it, another known card, so we can find it later. When we find the known card, we know that the spectator's selected card is next to it. In this section, we'll talk about a concept called the "key card."

Basic Key Card Control

A key card is a card that you place next to a spectator's unknown selected card. As an example, let's say that the jack of diamonds is your key card. When you have the spectator return his card to the deck, you put your jack of diamonds on top of the selected card. To later find the spectator's card, all you have to do is find the jack of diamonds and the card underneath it will be the selected card.

First you have to create a key card. Some tricks require special cards that you always use as a key card. Most card tricks rely on key cards that you designate on the fly, usually by looking at the top or bottom card of the deck. When you secretly look at and remember a card in the deck, magicians call this a "glimpse."

Here's a basic "pick-a-card/find-a-card" trick. This isn't a great trick at this point, but I'm using it to explain the concept. In Chapter 5, I'll show you how to use this skill to perform a superior trick. While technically, this is a trick, it's not one that you would probably want to subject your friends to.

Ask a spectator to take the deck of cards, mix them, and give them back to you. Ask the spectator to cut the deck and rest the upper half on the table. Tell the spectator to look at and memorize the top card that he cut to.

As the spectator picks up and looks at his selected card, casually lift up the edge of the deck that you're holding and look at the bottom card of the pack, as shown in Figure 4.1. To make this easier, you can turn away as if you're making sure that you can't see the selected card, but look at and memorize the bottom card. In this example, your key card would be the five of diamonds.

Figure 4.1
Casually glimpse at the bottom card of the pack as the spectator looks at his card.

Ask the spectator to return his card to the top half of the deck that's on the table and then replace the lower half back on top of the selected card. At this point, your key card is resting on top of the spectator's selected card.

To later find the card, all you have to do is look through the deck and find your key card and the card underneath it will be the spectator's card. As you'll see in this chapter, we'll also show you how to perform a cut that looks as if it's mixing the deck but is not, which keeps your key card in its spot next to the spectator's card.

If this method of creating a key card feels bold, it is. You learned how to perform the Hindu and overhand shuffles in Chapter 3; now we'll also teach some methods that will allow you to glimpse a card as you are seemingly mixing the cards.

Overhand Shuffle Key Card Glimpse

This method of glimpsing the bottom card, which becomes your key card, relies on the overhand shuffle that you learned in Chapter 3. You glimpse your key card as the spectator returns a selected card.

Ask a spectator to mix the cards and then hand the deck back to you. Spread the cards in your hands to allow a spectator to choose any card. You ask the spectator to look at and remember the card and show it to others so they may do the same.

As the spectator is showing her card to others, get the deck ready to perform an overhand shuffle.

When the spectator is ready to return her card, begin an overhand shuffle by removing the bottom half of the deck and sliding off a couple of portions of the deck into the left hand, just as you would if you were performing the regular shuffle.

Extend and turn the bottom half so the spectator may return her card on top of it. As the spectator returns the card, momentarily look at the bottom card of the pack that's held in your right hand. Note that you may have to turn your right hand a bit to view the bottom card of the pack, as shown in Figure 4.2. This is your key card. In this example, your key card would be the five of diamonds.

Figure 4.2
Creating a key card during an overhand shuffle, as the spectator returns a card.

Once the spectator replaces the card, put the entire pack that you're holding with your right hand onto the cards in your left hand. Your key card now resides on top of the selected card.

Hindu Shuffle Key Card Glimpse

This method of glimpsing the bottom card, which becomes your key card, relies on the Hindu shuffle that you learned in Chapter 3. You glimpse at your key card as the spectator returns a selected card.

Ask a spectator to mix the cards and then hand the deck back to you. Spread the cards in your hands to allow a spectator to choose any card. Ask the spectator to look at and remember the card and show it to others so they may do the same.

As the spectator is showing her card to others, get the deck ready to perform a Hindu shuffle.

When the spectator is ready to return her card, begin a Hindu shuffle by removing the bottom half of the deck and sliding off a couple of portions of the deck into the left hand, just as you would with a regular shuffle.

Extend the bottom half to the spectator so she may return the card on top of it. As the spectator returns the card, momentarily look at the bottom card of the pack that's held in your right hand, as shown in Figure 4.3. This is your key card.

Once the spectator replaces the card, put the entire pack of cards that you're holding with your right hand onto the cards in your left hand. Your key card now resides on top of the selected card.

Figure 4.3
Using a Hindu shuffle to view a key card as a spectator returns a card.

Charlier—One Hand Cut Glimpse

If you've mastered the one-hand Charlier cut, you can use the move to glimpse the bottom card. When you perform the cut, turn your hand to the left a bit and you'll notice that at the point where you're separating the decks and are beginning to raise the lower half to the top, you'll get a clear view of the card that will end up on the bottom, as shown in Figure 4.4.

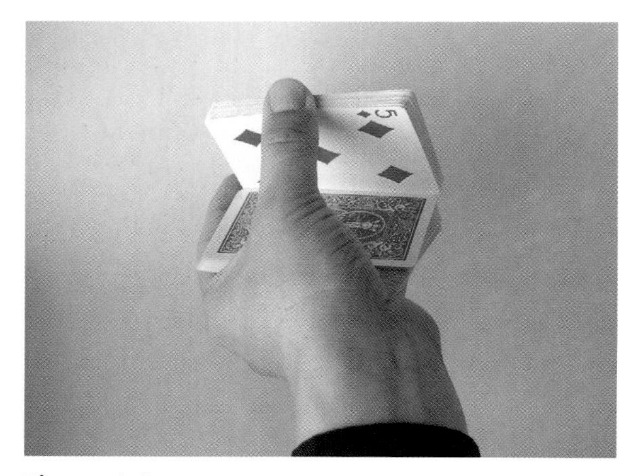

Figure 4.4
You can glimpse the bottom card as you perform a one-handed Charlier cut. This cut here is at the mid-point before you exchange the position of the halves.

This glimpse is very deceptive because you have every reason to look at your hands as you perform the cut.

Overhand and Hindu Shuffles—Bottom Card Glimpse

For some tricks, it's valuable to know the bottom card. You can use either the overhand or Hindu shuffles to accomplish this.

When using the overhand shuffle, you glimpse the card just before you place the last portion of the packet on the top of the deck. Perform your overhand shuffle as you normally would. When you are down to the last portion that you will throw on top of the deck, shift the pack in your left hand to the left a bit and slightly forward as you raise the last portion. This will give you a momentary glimpse of the bottom card, as shown in Figure 4.5.

Many of you will find this move awkward and will probably prefer the method that involves the Hindu shuffle that's explained next.

Figure 4.5
You can glimpse the bottom card as you perform an overhand shuffle.

When using a Hindu shuffle, glimpse the bottom card as you prepare to place the final portion of the packet in the right hand on top of the deck. You'll shift the packet in your left hand slightly forward to quickly view the bottom card, as shown in Figure 4.6.

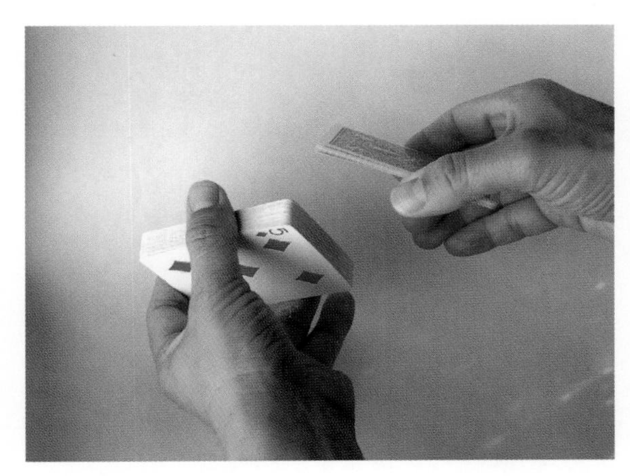

Figure 4.6
You can glimpse the bottom card as you perform a Hindu shuffle.

> **Hint**
>
> In both shuffles, it's natural to have a slight pause just before you throw on the final portion of the deck. Just try to accomplish the glimpse within the natural rhythm of your shuffle.

Hindu Versus Overhand Shuffle

As mentioned in Chapter 3, the overhand shuffle is commonly used, whereas the Hindu shuffle is only used by magicians, but is often easier to perform. I recommend that you use the shuffle that feels most comfortable for you. I often use a Hindu shuffle in my card magic and have never heard a negative comment from a spectator regarding it.

Holding and Controlling a Break

As mentioned earlier, magicians often control the position of a card in the deck. Usually, the objective is to allow a spectator to return a selected card anywhere, but quickly bring the card to the top or bottom of the deck without anyone noticing. And in most card tricks that rely on such a control, the card is usually brought to the top of the deck.

The first thing that you have to learn is to control a card with a marker of sorts, your little finger. This is something that magicians call a "pinky break." A spectator's selected card is placed onto the bottom half of the deck and as the magician covers the card with the upper half of the deck, he inserts his little finger so that the tip rests on top of the selected card, which holds the position. This is known as a "break," which is shown in Figure 4.7.

Figure 4.7
The pinky break allows you hold a place in the deck where a spectator has returned a card.

Despite extending the pinky finger into the deck to hold a card's position, the front of the deck is pushed together so that it looks mostly normal, as shown in Figure 4.8.

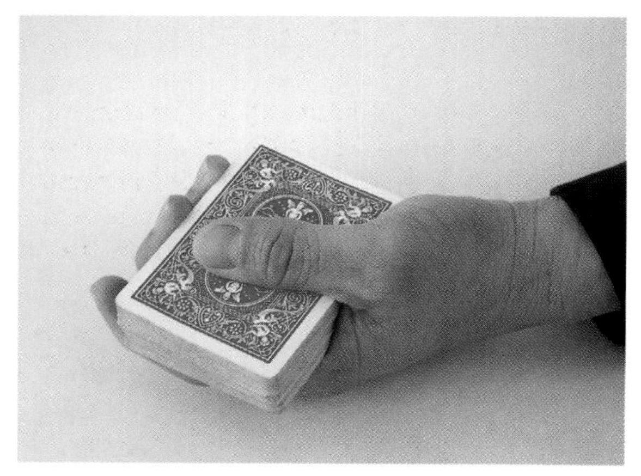

Figure 4.8
Even though you're holding a break, the deck looks mostly normal from the front.

The skill in holding a break is secretly sliding the tip of your pinky over the selected card at the last split second as you bring down the upper part of the deck—the movement of the upper deck provides cover for your pinky so spectators don't see the motion.

The second important aspect is closing the front of the deck so that it looks normal, at least as normal as a deck with a pinky finger in it can look. You'll have to check the position of your pinky and its relation to the cards to determine the optimum positioning.

You're now holding a break over the selected card. Now you have to bring the card to the top of the deck. For this, we'll teach a move known as a double-undercut.

Double-Undercut

With your pinky firmly holding a break in the deck, the best of the easiest techniques to bring the card to the top is the double-undercut, which is often used by pros. With a double undercut, you're simply cutting the selected card to the top of the deck, by using two cuts, thus the name. You can perform a double-undercut on the table or in the hands. We'll cover both.

If you're not sure if you'll have a table to work with, it's better to learn the double-undercut in the hands so you can use it whether or not you have a table nearby. The double-undercut in the hands is harder to learn and takes more practice, but it's well worth the effort.

Double-Undercut—Table Method

Hold the position of a selected card by using a pinky break. What you're going to do now is break the upper portion of the deck above your finger break into two sections and place these sections on the table, one on top of the other.

Mentally divide the upper half of the deck above the pinky into parts, and using the right hand, pick up the upper portion of the top half and place it on the table, as shown in Figure 4.9.

Now take the portion of the upper half that is resting on top of your pinky and lift it with the right hand and stack it on top of the half that's already on the table, as shown in Figure 4.10. When you remove the second half, be sure to take your pinky off of the top of the bottom half.

Figure 4.9
Remove half of the deck that's above your break and place it on the table.

Figure 4.10
Remove the second section above the break and stack it on top of the pile on the table.

Finally, place the remaining portion that has been sitting under your pinky onto the pile. The spectator's selected card is now on top of the deck.

Double-Undercut—In the Hands

Hold the position of a selected card by using a pinky break. What you're going to do now is break the lower portion of the deck below your finger break into two sections and place these sections on top of the deck that's held in your hand.

To perform a double-undercut, you'll be transferring the deck into your right hand and holding it in a position known as the "Biddle grip," as shown in Figure 4.11. Notice how the first finger curls on top of the deck and how the deck is held by the thumb and fingers.

Figure 4.11
The Biddle grip.

While placing the deck into the right hand isn't hard, you'll also have to use the right thumb to hold the break, as shown in Figure 4.12.

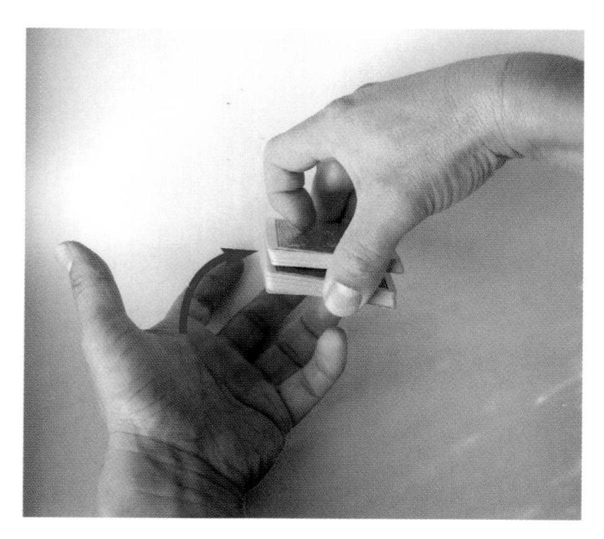

Figure 4.12
The thumb holds the break given to it by the pinky.

You're actually transferring the break from the left pinky to the right thumb. The thumb doesn't actually go into the break the way the pinky does, it simply holds a gap. This is going to take some practice.

To begin the double-undercut, get a pinky break to hold the selected card's position. Transfer the deck and its break to the right hand, which holds the deck in a Biddle grip. The right thumb now holds the break.

Mentally divide the lower half of the deck below the (thumb) break into two parts and using the left hand, pick up the lower portion of the bottom half and place it on top of the deck, as shown in Figure 4.13. You may need to turn the cards slightly so they can pass through your fingers and be laid on top of the deck.

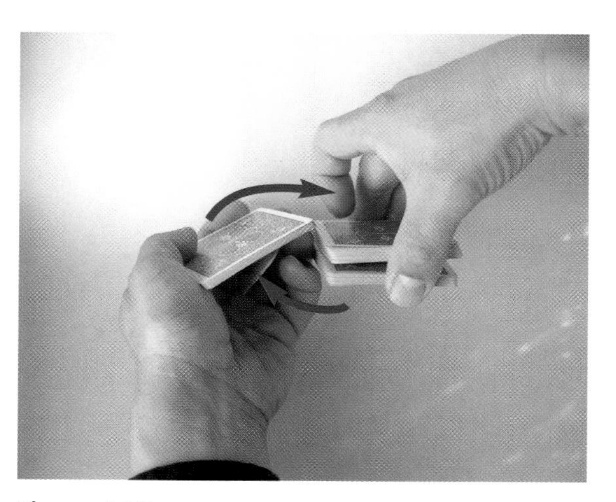

Figure 4.13
Transfer the lower half below the thumb break from the bottom to the top of the deck.

Now, take the portion of the lower half that is resting below the break and lift it with the left hand and stack it on top of the deck, as shown in Figure 4.14.

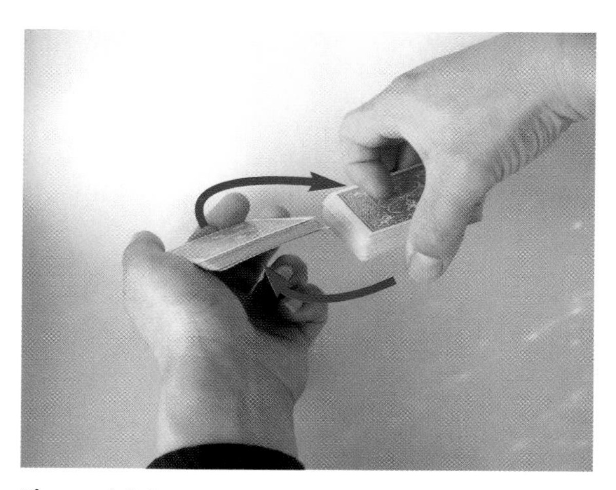

Figure 4.14
Take the remaining portion below the break and bring it to the top of the deck.

The spectator's selected card is now on top of the deck.

Hint

The hardest aspect of this version of the double-undercut is transferring the break from your left pinky to your right thumb. This is a good move to practice as you watch TV. You want to practice until you can confidently perform it without losing the break and without thinking about it. And if you don't want to think, what better way than to watch TV?

Another important aspect of performing sleight-of-hand is to perform the moves casually. If you appear nervous and are focused on your hands, spectators will suspect that you are doing "something." This is why it's extremely important to be able to perform the moves without thinking about them, and even talk with spectators as you execute the moves. They'll just think that you're mixing the cards when you're actually controlling the location of their card.

©istockphoto.com/Justin Horrocks

Overhand False Shuffle and Card Control

Here's a technique that looks as if you are performing an overhand shuffle to mix the cards, but you're actually retaining the order of the cards in the top portion of the deck. This technique also allows you to accept a chosen card from a spectator and appear to lose it, but cause the chosen card to end up on top of the deck. We'll teach both methods.

Retaining the Order of the Top Portion of the Deck

Begin to execute an overhand shuffle as described in Chapter 3. Slide the first portion of the deck into the left hand. This is the portion that will remain untouched and will end back up on top of the deck. The bigger the portion of the deck that you initially slide into the left hand, the more of the deck that will remain in its original order, as shown in Figure 4.15.

Figure 4.15

The first packet that you throw onto your hand will remain in order.

Hint

To make the shuffle easier and reduce the likelihood that the deck will get out of order, it helps to ensure that the first packet that you deal into your left hand is squared so the edges of the cards are aligned and relatively straight.

Slide a single card from the right hand into the left hand packet and then push the card towards yourself about half an inch, as shown in Figure 4.16. Magicians call this a "jog." And in this instance, it's an "in-jog" because the card is pushed towards you.

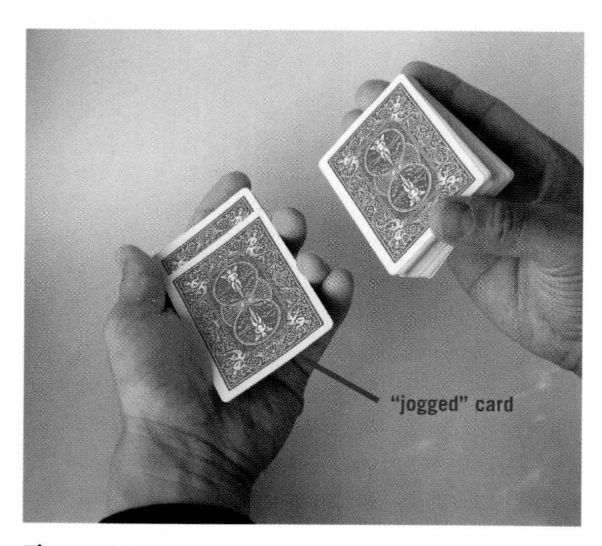

Figure 4.16
"Jog" a single card.

Shuffle the rest of the deck on top of the packet in the left hand as you would in a normal overhand shuffle. Be careful that you don't inadvertently shift or move the "jogged" card and lose your place in the deck. Your deck will look like that in Figure 4.17.

Figure 4.17
After performing an overhand shuffle, the position of the spectator's card is held with the jogged card.

As you become more confident with this shuffle, you can shuffle cards on top of the "jogged" card in a more haphazard manner than that depicted in Figure 4.17. By doing this, you help hide the "jogged" card.

Using the thumb of your right hand, push against the jogged card so you create a small break that you can hold with your thumb. Note that you are now holding the deck in a Biddle grip, but the face of the deck is towards your palm. This is shown in Figure 4.18.

While holding the break with your right thumb, lift a portion of the deck above the break and carry it to the right, as shown in Figure 4.19.

Shuffle the cards from the right hand to the left until you use up the cards up to the break. Throw the entire lower portion of the packet under the break, which is the original top portion of the deck, onto the top to finish the shuffle. You have maintained the order of the top of the deck.

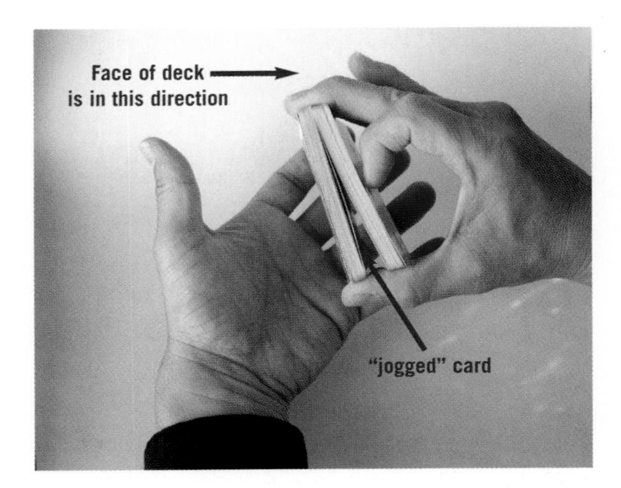

Figure 4.18

Press your thumb against the jogged card to create a small break.

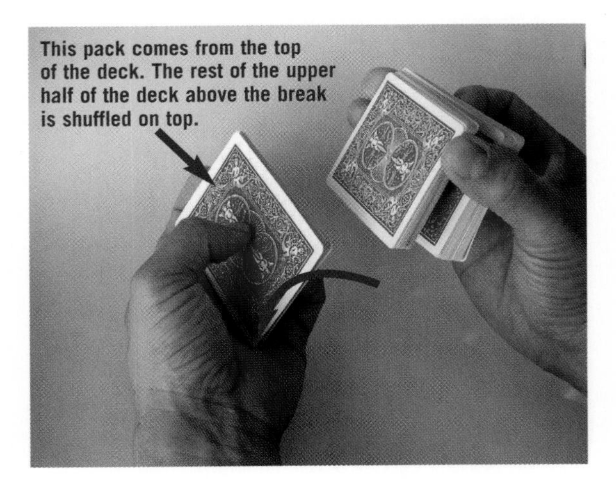

Figure 4.19

Hold the break with your thumb as you carry away part of the deck and then shuffle the deck to the break.

You'll have to practice this one a lot so it's smooth. In the beginning, you'll constantly lose the jogged card and gradually learn how to remedy this.

Controlling a Card with the Overhand False Shuffle

Begin an overhand shuffle and throw a few small portions into the left hand. Stop and ask the spectator to replace his card on top of the deck in the left hand. From here, slide the single card and jog it as explained earlier, create the break, and then perform the false overhand shuffle. The spectator's card will end up on top of the deck. To the spectator, it will appear as if you shuffled his card into the deck and lost it.

Hindu Shuffle and Card Control

Here's a technique that looks as if you are performing a Hindu shuffle to seemingly mix the cards, but you are actually retaining the order of the cards in the top portion of the deck. This technique also allows you to accept a chosen card from a spectator and appear to lose it in the deck, but cause the chosen card to end up on top. We'll teach both methods.

Execute a Hindu shuffle as described in Chapter 3. Slide the first portion of the deck into the left hand, as shown in Figure 4.20. This is the portion that will remain untouched and will end back up on top of the deck. The bigger the portion of the deck that you initially slide into the left hand, the more of the deck that will remain in its original order.

Using the right thumb and fingers, grasp the packet that you just placed into the left hand, and continue to hold the rest of the pack. Using the right thumb, create and maintain a break between the two halves of the deck, as shown in Figure 4.21. I like to hold the other side of the deck with my right second finger.

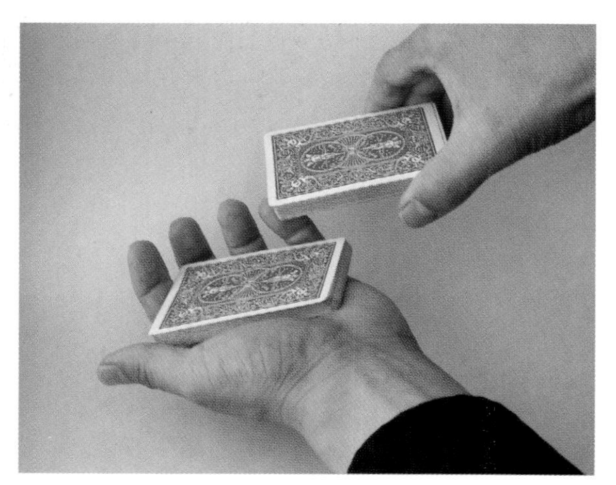

Figure 4.20
The first packet that you throw onto your hand will remain in order.

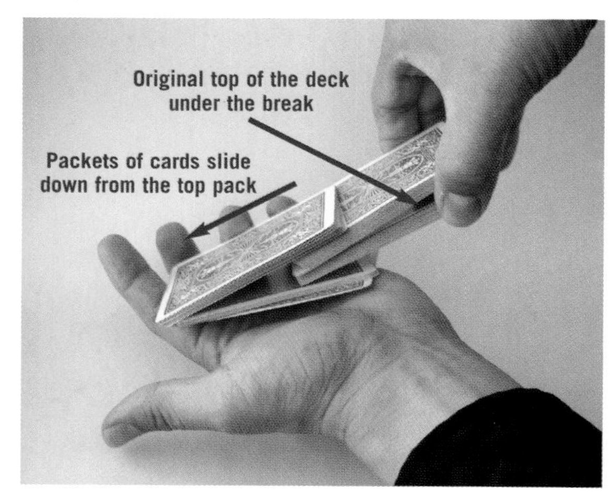

Figure 4.21
Maintain a break using the right thumb.

Carry away the bottom half and the upper half while shuffling a portion off of the top, just as you would in a normal Hindu shuffle. Continue to shuffle cards off of the top until you get to the break, as shown in Figure 4.22. Be careful that you don't lose the break.

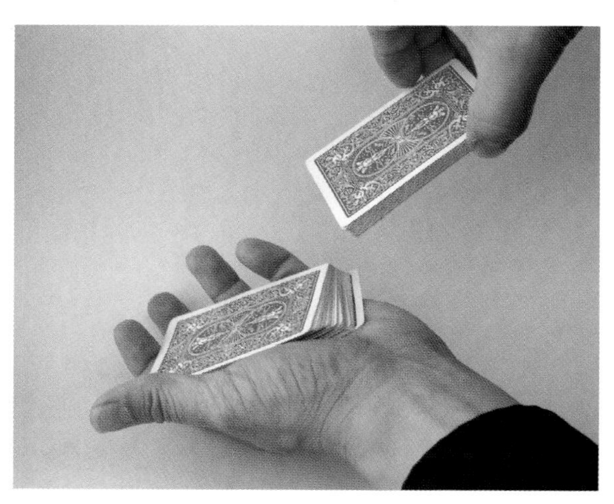

Figure 4.22
Shuffle cards until you reach the break.

> **Hint**
>
> To ensure that I don't lose my break, I often insert the tip of my right first finger into the break. This is why I recommended earlier that you hold the side of the deck with the right second finger.
>
> You'll have to practice this one a lot so it's smooth. In the beginning, you'll constantly lose the break and gradually learn how to remedy this.

Throw the last portion of the pack, the portion that was originally the top, onto the pack. You have maintained the order of the top portion of the deck.

Between the two false shuffles, the Hindu is the easier one to master and it's less likely to lose one's place in the deck. There's no need to learn both techniques. Simply try both and choose the one that feels most natural to you and that you're most comfortable performing.

Changing the Position of a Card

AT TIMES, YOU'LL WANT TO MOVE a card from one place in the deck to another, typically from the bottom to the top and the top to the bottom. To support the tricks in later chapters, here are moves that use the overhand and Hindu shuffles to transfer a card from the top to the bottom and vice-versa.

Overhand and Hindu—Top to Bottom

If you want to place the top card on the bottom of the deck, perform a standard overhand or Hindu shuffle and begin by shuffling a single (the top) card and then perform the standard shuffle. The top card is now on the bottom.

You can also bring a card from the bottom to the top via an overhand or Hindu shuffle. This takes a bit more practice. Perform your standard overhand or Hindu shuffle. As you exhaust the cards, just be sure to count the last few cards one at a time, which will place the card that was originally on the bottom on the top of the deck.

False Mixing

WITH THE USE OF A BREAK and a double-undercut, or an overhand or Hindu control, you now have the spectator's card resting on top of the deck. If you have done this correctly, the spectator should have no idea that the card is resting on top.

However, there are spectators who will question your every move and guess that a card is already on top of the deck (they may have read magic books and know this for a fact). For these spectators, we want to throw them off of the trail by making it look as if we are mixing the deck when we are not. In addition to our overhand and Hindu controls, we can also perform a false cut.

Another use for false mixing is to appear to mix up a deck when it's already been set in a particular order. This is particularly useful when performing some of the tricks in Chapter 2 that require that the deck be set a certain way at the start of the trick.

False Three-Way Cut—Version 1

This false cut breaks the deck into three sections and reassembles the deck so the order has not changed. This one is particularly good when you're working on a table, but you can, with practice, perform this one in the hands.

Hold the deck horizontal with a long side towards you. The first finger and thumb of the right hand are holding the lower third of the deck. The first finger and thumb of the left hand are holding the middle third of the deck, as shown in Figure 4.23. For illustration purposes, I have assigned the designations: (A), (B), and (C) to represent respectively, the top, middle, and lower thirds of the deck.

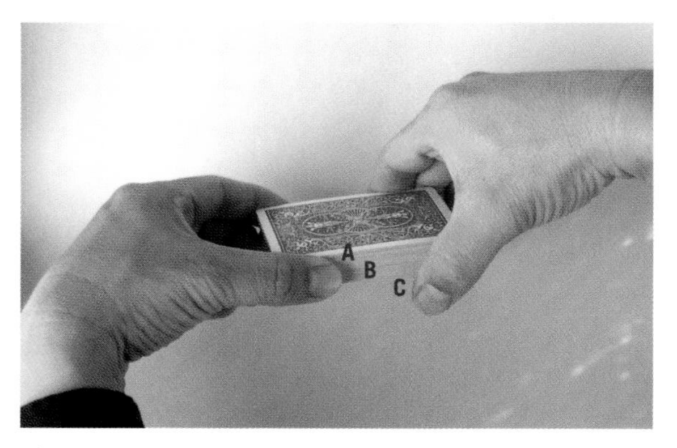

Figure 4.23
Holding the deck for the three-way false cut.

Using the first finger and thumb of the right hand, pull off and separate the lower third of the deck and carry it to the right and then on top of the deck. The lower third (C) ends up resting near the top of the fingers of the left hand at an angle, as shown in Figure 4.24.

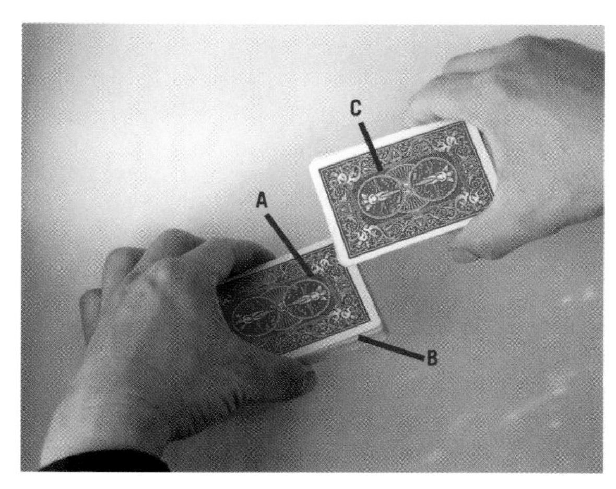

Figure 4.24
Bring the lower third (C) up to the top.

Using the second finger and the thumb of the right hand, pick up and separate the top third of the deck (A) and carry it to the right. At this point, the top third of the deck (A) is held under the lower third of the deck (C) in the right hand. The left hand maintains its grip on the middle third of the deck (B), as shown in Figure 4.25.

Figure 4.25
Pick up the top part of the deck (A) with the right hand.

The first finger and thumb of the right hand drop the lower third (C) onto the table, as shown in Figure 4.26. You're dropping portion (C) and quickly taking (A) out of the way. To make this easier, it helps if you hold the lower third (C) as far forward as you can, and the top portion (A) as far back as you can.

The left hand drops the middle third (B) onto the deck (C) on the table, as shown in Figure 4.27.

The right hand drops the remaining third (A) onto the deck on the table.

You have cut the deck into three sections and reassembled them back into their original order.

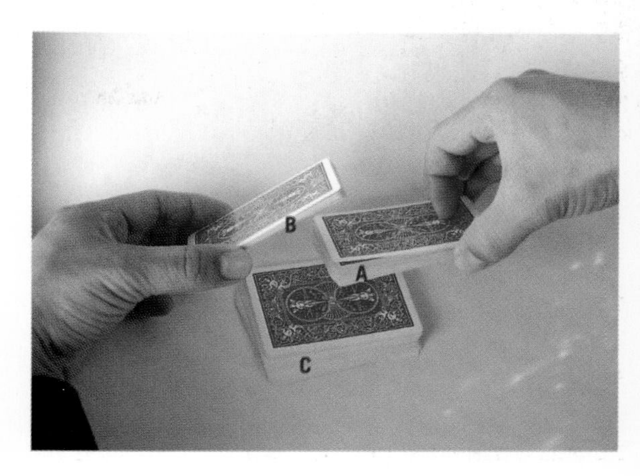

Figure 4.26
Drop (C) on the table.

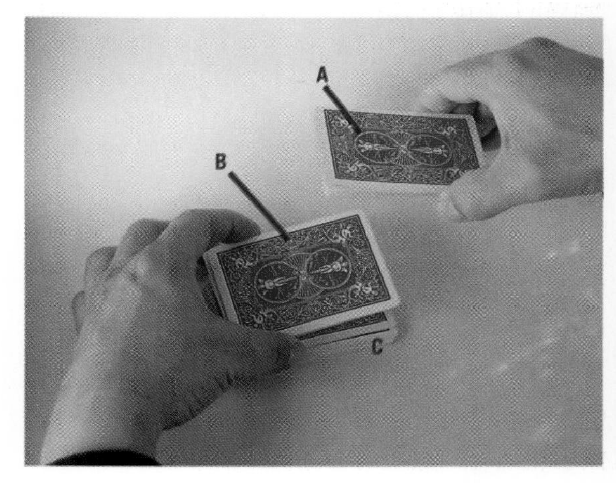

Figure 4.27
Drop (B) onto (C).

If you want to perform this false cut in your hands without a table, you have to practice using your left hand to hold and reassemble the deck portions as they are coming back together.

False Three-Way Cut—Version 2

Like the first false cut, this move breaks the deck into three sections and reassembles the deck so the order has not changed. This cut happens in the hands and does not require a table.

Hold the deck with a short end towards you and the deck on its side, as in Figure 4.28. The second finger and thumb of the right hand are holding two-thirds of the deck (B/C) and the right first finger is curled on bottom of the deck. The left hand fingers are holding the deck on its side cradling the edge along the first knuckle. The left thumb is resting on the upper edge.

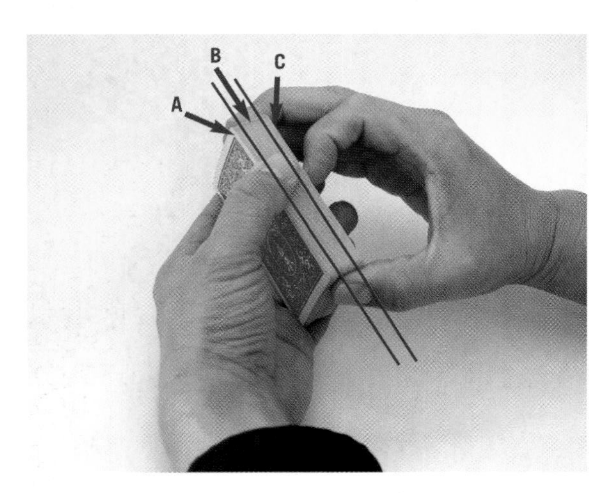

Figure 4.28
The grip for the false three-way cut.

The right second finger and thumb pull and separate two thirds of the deck (B/C) and carry it up and to the right, as shown in Figure 4.29.

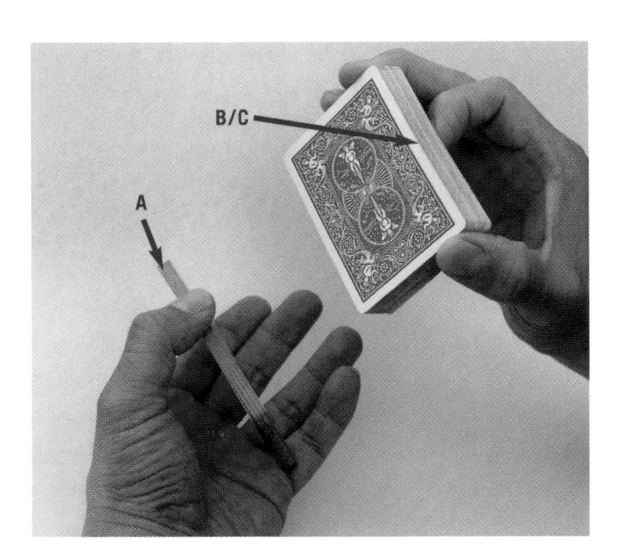

Figure 4.29
Lift sections B and C.

At the same time, the left thumb releases its hold on its third (A) and allows it to pivot and rest against the tips of the left fingers, as shown in Figure 4.30.

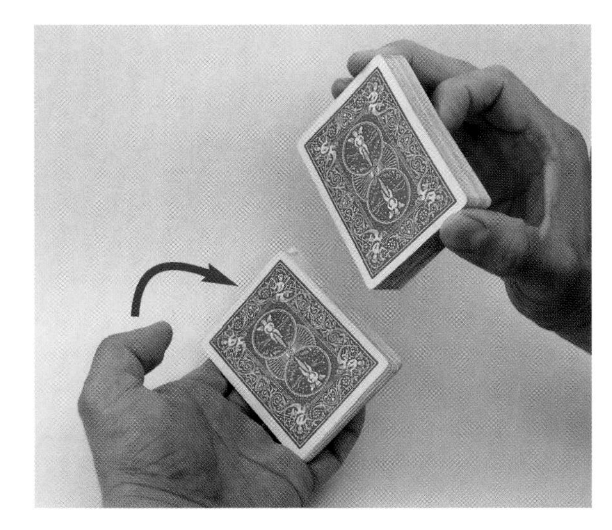

Figure 4.30
Allow (A) to pivot and rest against the left fingertips.

The right hand drops its section (B/C) into the palm of the left hand to the left of (A), as shown in Figure 4.31.

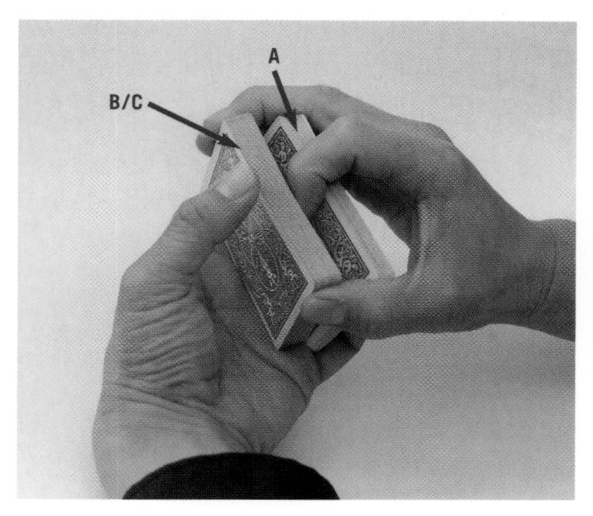

Figure 4.31
The right hand drops (B/C) into the left hand.

The left thumb grips half of the section (B/C) and divides it and pulls away (B). Meanwhile, the first knuckle of the right second finger and the right thumb grip (C), as shown in Figure 4.32. The deck is now in three parts. This is the most difficult step in the cut.

The right hand carries away sections (C) and (A). Notice that the right hand maintains a break between the two sections, as in Figure 4.33.

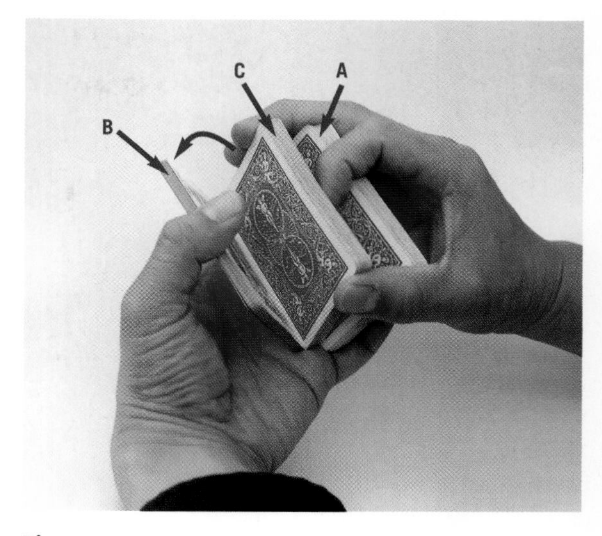

Figure 4.32
(B) is broken away from (C).

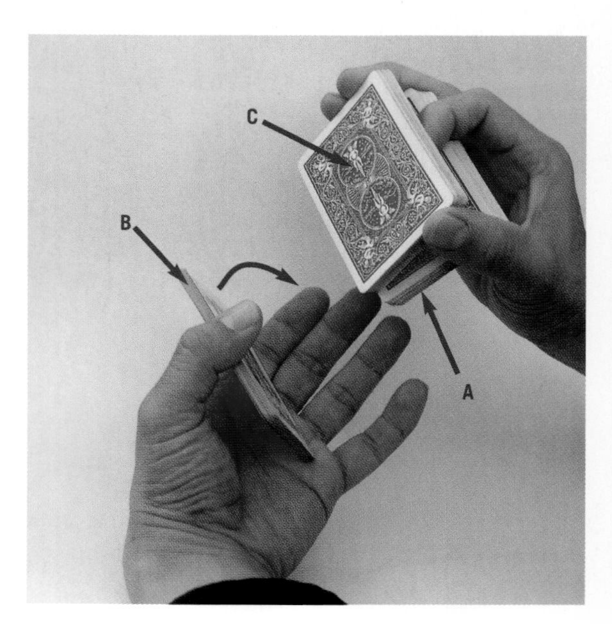

Figure 4.33
The right hand carries away (C) and (A).

The left hand allows the thumb to release its hold on section (B) and push the card so that it pivots and rests against the tips of the left fingers, as in Figure 4.34.

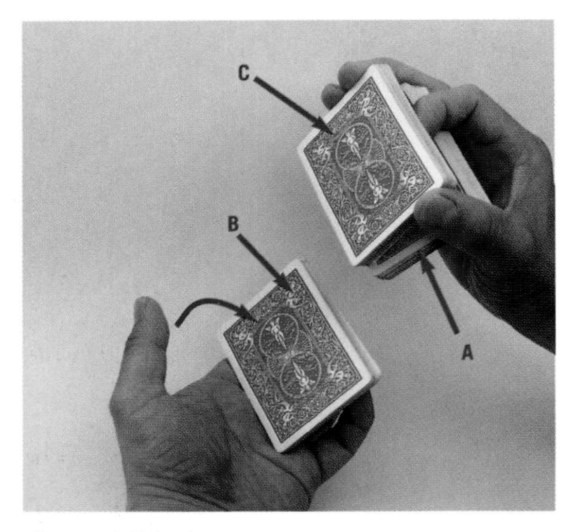

Figure 4.34
(B) pivots in the left hand.

The right hand drops (A) onto (B), as shown in Figure 4.35.

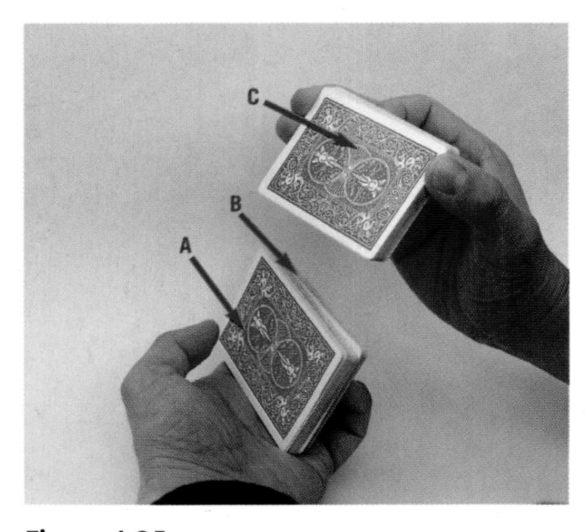

Figure 4.35
Drop (A) onto (B).

The left hand uses its fingers to push (A/B) so it pivots back against the thumb, as shown in Figure 4.36.

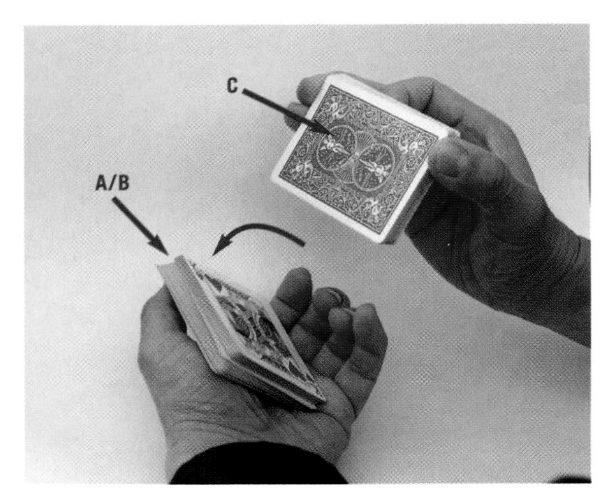

Figure 4.36
(A/B) pivots back.

The right hand drops the uppermost third (C) on top of the deck. You have cut the deck into three sections and reassembled them back into their original order.

Maintaining the Top Card with a Riffle Shuffle

If you want to retain a card on the top of the deck, it's not difficult to perform a riffle shuffle and keep a card at the top. You can use your choice of a table or riffle shuffle and a bridge, as taught in Chapter 3.

All you need to do is make sure that the top card is the last card to fall on top of the deck as you riffle shuffle cards. This will also depend on how you split the deck to perform a riffle shuffle.

I suggest that you reverse the top card of the deck so it's face up and then execute your normal split of the deck and determine where the top (face-up card) ends up. If it's in the left hand, you'll want the top card in the left hand to be the last to fall. If it's in the right hand, you'll want the top card in the right hand to be the last to fall.

Try and be consistent with your shuffle so you don't have to think about where the top card is. Simply execute the shuffle the same way each time.

Once a card is returned to the deck, a riffle shuffle that keeps the selected card on top followed by a false cut should convince spectators that their card is lost in the deck.

Free Choice?

THE THIRD MAJOR CARD SLEIGHT that magicians often use is allowing a spectator to freely select a card when the magician has completely controlled the card that they pick. Magicians call this technique a "force."

We offer several forces here that you can learn and use. Pick one that feels comfortable for you.

The Cross-cut

I've seen pros use this very simple force. This one relies on misdirection and you'll need to have some good patter to momentarily take the attention away from the deck. But this one works.

Figure 4.37
The cross-cut force.

In the cross-cut, the card that you want spectators to select is on top of the deck. You can place the card on the top of the deck before hand so your card is already in place when you begin a trick. If you like, you can perform a riffle shuffle that maintains the top card and then a false cut.

Rest the deck on the table and ask the spectator to cut the deck but to not complete it.

Lift up the bottom half and place it on top of the upper half that is resting on the table, but turn the deck 90 degrees, as shown in Figure 4.37.

At this point, you'll want to talk a bit and take the attention away from the cards. Here you can talk about the number of cards in the deck, 52, and the statistical probability of selecting a particular card, or any other theme that suits the trick and your personality.

After talking a bit, you simply lift off the top half of the deck and tell the spectator to look at the card that's on top, as shown in Figure 4.38. They'll be looking at the card that was originally on top of the deck. At this point you ask the spectator to look at and remember the card and show it to others.

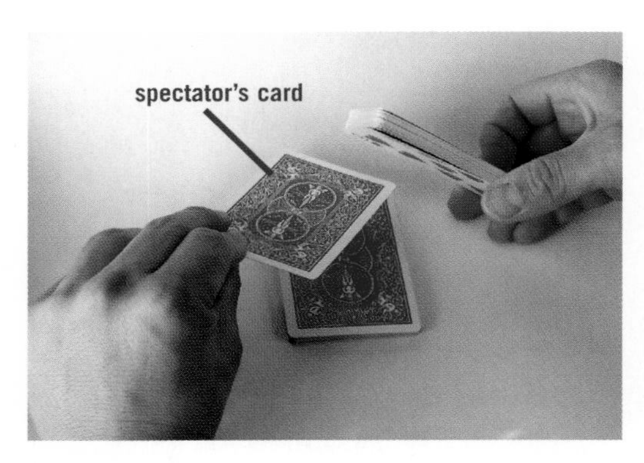

Figure 4.38
With the cross-cut force, the spectator selects the original top card.

At this stage of the game, you already know the identity of the card, which you can use to your advantage. Another plus, you can have the spectator return the card to the deck and freely mix it to their satisfaction. You don't need to control the card because you already know it.

The Hindu Force

This move is easy and quite deceptive. In the Hindu force, the card that you want spectators to select is on the bottom of the deck, as shown in Figure 4.39.

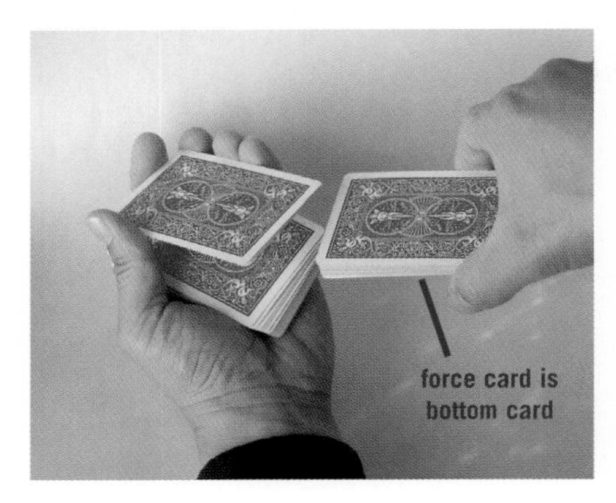

Figure 4.39
The Hindu force makes the spectator pick the bottom card.

Simply execute a standard Hindu shuffle and ask the spectator to say "stop."

When the spectator says "stop," show the bottom card that is held in the right hand, as in Figure 4.40.

Figure 4.40
Turn the deck over when a spectator says "stop."

> **Tip**
>
> This move is more effective when there's a slight pause between when you stop shuffling and when you turn over the stack in your right hand to reveal the bottom card. You don't need to pause long, just a half second. I like to pretend that I'm straightening the pack in the left hand by gently shaking it side to side a bit. It's just a little nuance that momentarily takes the spectator's attention away from the deck and the shuffle.

You can use the Hindu force to show a card and then immediately mix the card back into the deck so you shouldn't know the identity of the card. With the Hindu force, it's rather clumsy to give the selected card to the spectator so they can show it around and freely return it to the deck.

Deeper Cut Force

Here's an easy force that uses two cuts. Place your force card on the top of the deck and walk the spectator through the following steps.

Ask the spectator to cut a small portion off of the top of the deck and turn this portion face up on top of the deck, as shown in Figure 4.41.

Figure 4.41

The spectator cuts a small portion of cards from the top and turns it over.

Ask the spectator to reach further into the deck and turn over more of the deck, including the face-up portion turned over earlier, as in Figure 4.42.

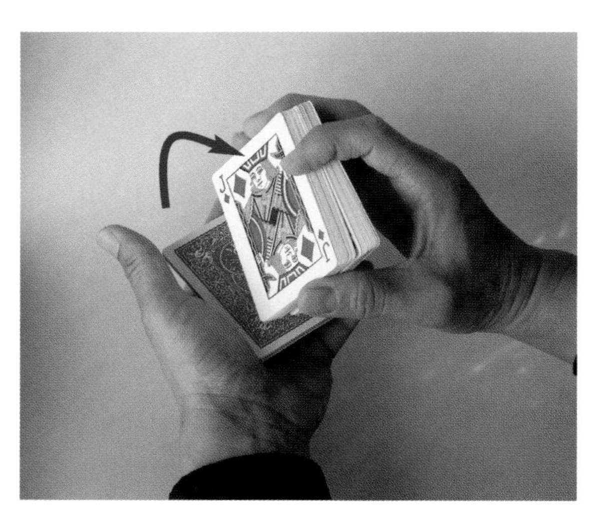

Figure 4.42

The spectator cuts again and turns the packet over.

Ask the spectator to thumb down through the cards from the top and find the first face-down card, as shown in Figure 4.43. Ask the spectator to take the card and remember it. The card will be the force card that was originally on top of the deck.

Figure 4.43

The spectator turns over and remembers this card.

Flourishes

MORE CARD JUGGLING THAN CARD magic, a good card flourish will dress up your routines. Here are a couple of flashy moves.

The Card Fan

The card fan, as shown in Figure 4.44, is practically the mark of a magician. You essentially spread the cards in an even manner as the deck is held in your hands.

Figure 4.44
A card fan.

This one is difficult to explain but once you have the proper hand position, you'll need to practice the move until you get the pressure just right. By the way, it helps to have relatively new cards to make a fan. If you have dirty cards or plastic-coated cards, it's just about impossible to fan them.

Hold the deck between the left thumb and the left fingers as shown in Figure 4.45. Notice that the back of the deck is being held with the left first, second, and third fingers. The corner of the deck that's being held by your thumb and mostly by your second finger on the other side, will act as a pivot point.

Figure 4.45
Holding the deck in preparation for a card fan.

Take your right thumb and press lightly forward against the edge of the deck and then push the cards in a circular motion to spread the cards. This is shown in Figure 4.46.

Figure 4.46
Push the cards in a circular motion.

That's about it. The rest is just practice. Experiment with the pressure exerted by the right thumb until you get something that resembles a fan. Once you begin to create a rough fan, you can refine the motion until you get even fans each time.

Card Springing

This is that fancy move where the magician takes a deck of cards and makes them fly from one hand to the other in a flashy manner. You'll receive lots of attention with this one.

Hold the deck in your right hand with your thumb on the bottom edge and your fingers on the top, as shown in Figure 4.47.

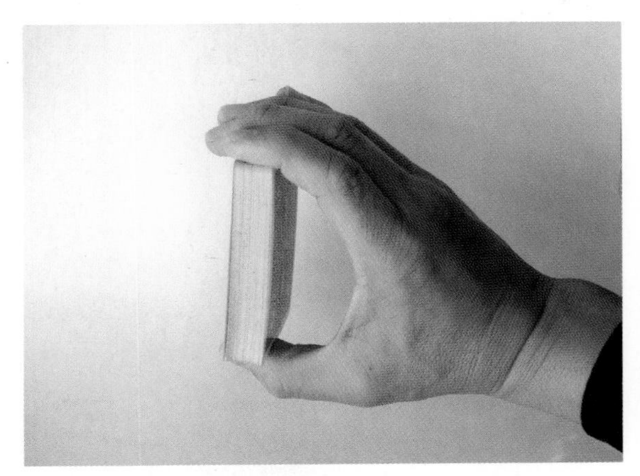

Figure 4.47
Holding the cards in preparation for a card spring.

By pressing the fingers down, you cause the deck of cards to bow in your hands. You'll want the cards to bow towards your palm, as in Figure 4.48.

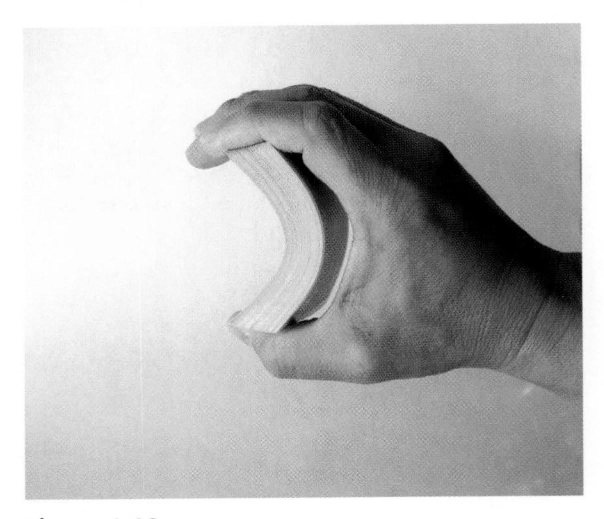

Figure 4.48
Bowing the cards

Your left hand has to catch the cards. For this, you hold your hand in the shape of a basket, as shown in Figure 4.49.

Figure 4.49
The left hand is prepared to catch the cards as the cards spring out of your hand.

You have a choice at this point to cause the cards to run off of your thumb or off of your fingers. Personally, I run the cards off of my thumb. For explanation purposes, we'll assume that you are running cards off of the thumb. If it's easier for you to run the cards off of your fingers, just adjust the instructions.

By pressing down on the deck, you cause the cards to slip off of your thumb, one-by-one. The cards will noisily spring out of your hand. With practice, you'll be able to spring cards a foot or even more.

Note how the cards are coming off of the thumb and how the left hand is held slightly below to catch the cards at the end of their trajectory.

Next Steps

With your newfound sleight-of-hand skills, you're ready to perform some great card magic that's explained in Chapter 5.

5

Great Card Tricks

NOW THAT YOU LEARNED SOME CARD handling in Chapter 3 and basic sleight-of-hand skills in Chapter 4, we can teach a whole new level of card tricks for you to perform.

Ready to try out your newfound sleight-of-hand skills? In this chapter, I'll teach you a variety of card tricks that will put your moves to good use.

Visualize the Tricks

BEFORE YOU READ THE DESCRIPTION and secret of each trick, you may find it helpful to view the performances of them on the accompanying DVD. Actually, I recommend that you watch the performance of each trick before you read its description and secret. For this, please refer to the section on "Card Tricks" that demonstrates each effect.

Finding a Spectator's Card (Two Themes)

Here's a couple of simple "find a card" tricks that will get you started. I've tried to create a couple of themes that you can use to dress up the effect. Both will allow you to interject some fun into the proceedings.

Effect

A spectator freely selects a card and returns it to the deck. You find the card by looking for a subconscious, subliminal, physiological signal from the spectator, or examine the cards looking for a fingerprint.

Materials and Requirements

A deck of cards.

You'll need to perform this one on a table.

If you want to apply the fingerprint theme, a magnifying glass adds to the fun.

Secret

Use of a key card.

Skills

The ability to use a key card, as explained in Chapter 4.

False cut (optional).

Performing the Trick

Ask a spectator to select a card and then return it to the deck. While you're free to use any method with a key card, I recommend the overhand or Hindu method of establishing and viewing the key card and then having the card returned to the deck with the key card on top of it. If you like, you can perform a false cut.

> **Reality Check**
>
> Your key card resides on top of the spectator's selected card.

The Psychological Theme

Tell the spectator that you will find his card by looking for a subconscious signal as you deal cards face up onto the table. Hold the deck face down in your left hand and turn over each card and place it onto a pile on the table. What you are looking for is your key card. When you see your key card, the very next card will be the spectator's chosen card.

Continue dealing until you've spotted the key card. Turn over the next card and announce that it is the spectator's card.

> **Tip**
>
> Feel free to dress this one up with some acting. You can pretend that you are passing the card and getting ready to turn over the next card, just as if you missed the spectator's card. And when you're getting ready to turn over the next card, look up at the spectator and then realize that you missed the signal and then point to the face-up card and identify it as the chosen one.

Don't forget as you deal to continually look at the spectator. After all, you're supposed to be looking for some signal from them that indicates the chosen card.

The Fingerprint Theme

If you're looking for a fingerprint, first examine the spectator's fingertip and tell him that you will retain an image of his print in your memory to compare against. A magnifying glass here offers some fun, comedic possibilities.

From the face-down deck, again, turn over and carefully examine each card. When you see your key card, you know that the next card is the spectator's. You can examine the spectator's card and reveal it to be the chosen card. As with the first theme, you can pretend to fail at recognizing the spectator's card and then come back to the card when you realize that you've missed something and then reveal it.

Spelling to a Selected Card

In this effect, you spell, one card per letter, to a spectator's selected card.

Effect

A spectator selects a card that is replaced and lost in the deck. You display the cards and try to identify the card through a nonverbal, subconscious cue. You fail to find the card and then ask the spectator to name his card. You then spell the name of the card and count down from the top of the deck—one card for each letter. When the chosen card is spelled, the card at that location is the selected card.

Materials and Requirements

A deck of cards. This one can be done at any time and there is no preparation to the deck. You will need a table or you can deal cards into a spectator's hands.

Secret

Use of a key card.

Skills

The ability to use a key card, as explained in Chapter 4.

False cut (optional).

Preparation

This one can be done at any time and there is no preparation to the deck.

Performing the Trick

Ask the spectator to mix the deck of cards. Take back the deck and spread the cards so the spectator may freely choose one.

When you have the spectator return his card to the deck, use an overhand or Hindu shuffle and glimpse the bottom card. This is your key card that you'll have to remember. Place your key card on top of the selected card, as explained in Chapter 4. For purposes here, we'll assume that the spectator's card is the three of clubs.

Comment on the fact that you've been studying nonverbal and subconscious cues and so to find the card, you're going to show the cards to the spectator and look for a signal from them, a reaction that they won't even know that they are doing.

Spread the cards face-up and begin to look for your key card. When you spot your key card, you can identify the spectator's card because it will be the one directly before your key card, as shown in Figure 5.1. Remember to periodically look at your spectator because you're looking for a nonverbal sign.

Figure 5.1
The spectator's card is directly before your key card.

Note the spectator's card and begin to spell its name. The spectator's card will be the first letter, your key card the second and so on. Keep spreading through the cards and spell the card in your mind, as shown in Figure 5.2. In the example, we spell "two of clubs."

Figure 5.2
Mentally spell the spectator's card as you spread through the cards.

Hint

The diamonds tend to make for a lot of spelling. You can probably remove the "of" that you would normally include in a card's name to save time.

When you've finished spelling the card, create a break using the first finger of the right hand. Just as you used a pinky break to hold a place in a deck that's held in dealing position in your hand, you'll use your first finger to hold a place in the spread of cards, as shown in Figure 5.3.

Figure 5.3
Hold a break with your first finger after you've spelled the card.

Keep spreading the cards and periodically looking at your spectator. You can have some fun with this as the spectator typically puts on his best poker face and will try and limit his expressions.

When you've spread through all of the cards, simply close the spread and cut the deck at the break, bringing the bottom half of the deck to the top. This will bring the first card that you will spell to the top of the deck. To cover this, you can simply ask the spectator if he saw his card because you weren't sure if you caught the signal.

Reality Check

The spectator's card resides a certain number of cards from the top that coincides with the number of letters in the card's name.

At this point, you can perform a false cut that maintains the order of the entire deck if you would like.

Ask the spectator to name his card. When the spectator does, tell your audience that you are going to spell the card's name.

Deal a card off of the top of the deck for each letter of the name. When you get to the last letter, turn over the card to reveal that it's the spectator's card.

Tip

You don't have to spell a card's name. Feel free to set up and spell other things such as the spectator's name, the city where you live, and so on.

Do as I Do

I like to present this trick as a laboratory experiment. Both you and a spectator perform the same operations on your respective decks, cutting and shuffling, and amazingly, you both choose the very same card.

Effect

You take a blue deck and the spectator takes a red deck. You perform the same operations to your respective decks, shuffling and cutting them, and you both select cards. In the end, you show that you and the spectator selected the same cards.

Materials and Requirements

Two decks of cards. Two different decks (distinctly different backs) make for a stronger presentation.

Secret

You secretly glimpse the bottom card of your deck before you exchange decks with your spectator. When the deck is returned, you look for your key card and the spectator's card will be right next to it.

Skills

The ability to use a key card, as explained in Chapter 4.

False cut (optional).

Performing the Trick

Bring out the two decks of cards and allow a spectator to select one. Ask the spectator to do exactly what you do to try and create a similar outcome.

You're free to cut and shuffle the pack. Just convince the spectators that the decks are thoroughly mixed.

Perform an overhand or Hindu shuffle and ask the spectator to do the same. As you complete the shuffle, glimpse at the bottom card of your deck, as shown in Figure 5.4. This is your key card. For example purposes, we'll use the five of diamonds.

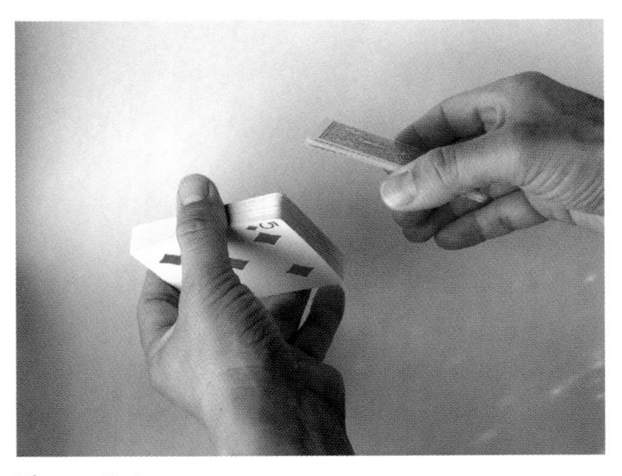

Figure 5.4

Use your favorite method to glimpse and remember the bottom card. Here I've used a glimpse via a Hindu shuffle, as taught in Chapter 4.

Trade decks with the spectator and set your deck on the table. Ask your spectator to "mirror" you as you cut your deck and set down the upper half to your right. (The spectator mirrors your movements cutting the deck and setting the upper half to his left.)

Pick up the top card and look at it, but don't memorize it. Ask the spectator to memorize his card. Keep repeating your key card silently to yourself ("five of diamonds," "five of diamonds"). The card you are now looking at is completely irrelevant. Make sure that other spectators don't see your card.

Figure 5.5
Cut the deck and look at the top card.

Place your card on the top of the upper half that's to your right, as shown in Figure 5.5. The spectator mirrors you.

Complete the cut with the bottom half of the deck.

Cut the deck once and then trade decks.

> **Reality Check**
>
> Your key card, the five of diamonds, now resides on top of the spectator's chosen card in your deck. Despite the cut, the order of the deck has not changed.

At this point, you can cut the deck as many times as you wish. But don't shuffle the deck as this will change the order of the cards.

Tell the spectator to look through his deck for his card and you will do the same with your deck. Look for your key card, the five of diamonds. The card below it will be the spectator's card.

By the way, if your key card is on the bottom of the deck, this means that the top card is the spectator's chosen card. Bring this card out and place it on the table face down. Ask the spectator to place his card face down next to your card.

Emphasize that each of you performed the same processes with your respective decks of card and thus, should have similar outcomes. Have the spectator turn over his card at the same time that you do. You'll show that they match.

The Rising Card

This is a classic of magic. We'll provide you with two routines. The second is more involved but will garner larger responses.

Effect

A selected card is returned to the deck and then placed into a card box. The magician holds the card box in his hand and the spectator's selected card mysteriously rises on its own accord out of the box.

Materials and Requirements

A deck of a cards and a doctored card box (see below).

Secret

There's a lot involved in knowing the spectator's card and controlling it to the correct position. But the rising part of the trick relies on a hole that's cut in the back of the card case. Using your finger, you slide the back (selected) card up and out of the top of the box.

Skills

Controlling a card to the top of the deck.

A false cut.

For the second version, forcing a card.

Preparation of the Box

You'll need to take some scissors or a pocket knife and carefully cut a hole in the back of the card box, as shown in Figure 5.6. The hole has to be big enough so your first finger can easily press against the back card and you have to allow space for the first finger to reposition itself to make the card continually rise.

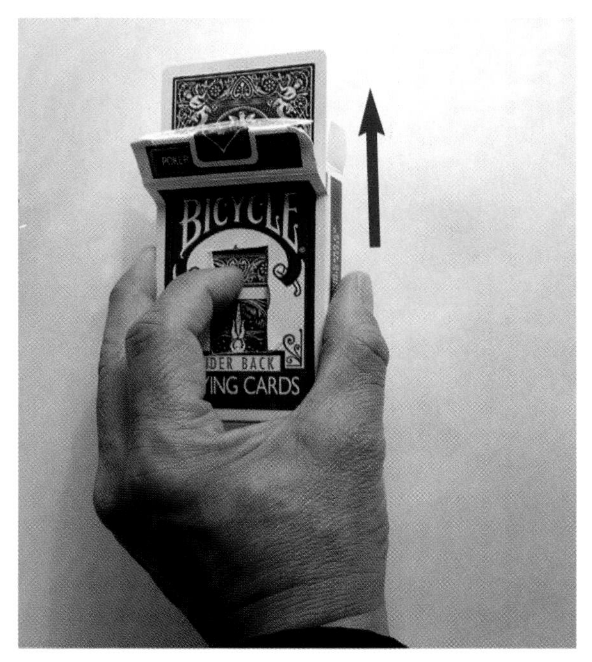

Figure 5.6
Cut a hole in the back of the card box.

In the first method, you have a spectator select a card and then control it to the top. The deck is placed into the card box and you make the selected card rise.

In the second method, the spectator selects a card and holds onto it. You place the deck into the card box and the mate to the spectator's card rises from the box. So if the spectator's card is the seven of diamonds, for example, the other red seven, the seven of hearts, will rise from the box.

By the way, when you make the card rise, it's important that you continually move the card box in a circular motion as the card rises. Otherwise, spectators may notice your finger moving and suspect something. The circular motion masks the mechanics, your finger, behind the rising card.

Rising Card: Method 1

Bring out the deck and ask a spectator to freely choose a card. Cut the deck and have the spectator return his card to the bottom half that's held in your hand.

Replace the top part of the deck and hold a pinky break.

Perform a double-undercut to bring the selected card to the top of the deck. If you like, you can control the card to the top either using the overhand or Hindu shuffle.

Place the deck into its card box.

Ask the spectator to name his card and then make the card rise from the box by pushing your finger against the card.

Rising Card: Method 2

This version reveals the mate to a spectator's selected card, even before he knows what his card is.

Beforehand, run through the deck and take out a pair of mates. For this example, we'll use the red sevens: the seven of diamonds and seven of hearts. Place these two cards face-down on top of the deck. We'll assume that the seven of diamonds is the top card.

Bring out the deck and ask the spectator to cut the deck. Perform the cross cut force and allow the spectator to pick up his card, which will be the seven of diamonds. Replace the original top half, which brings the mate of the selected card, the seven of hearts, to the top of the deck.

Place the deck into its card box and perform the card rise. Verify that it is the mate to the spectator's card.

Hint

The most dramatic way to perform this trick is to have the spectator not look at his selected card and set it on the table face-down. After you've performed the card rise, ask the spectator to turn over his card to reveal that it is the mate to the risen card.

This is now a different trick. You have not merely identified the spectator's card or simply found it, but the mysterious card rise has effectively named the spectator's card. Spectators will be astonished when the card rises and when they discover it's the mate to their own—you have created two magical moments.

Impromptu Prediction

Here's a trick where you appear to predict two cards that a spectator cut to. The only other solution that spectators will probably come up with is that you used a marked deck. If you perform this trick with a spectator's deck, it's a real fooler.

Effect

The spectator freely cuts a deck into two piles. The magician states the name of the card before he picks up each card on the top of each pile.

Materials

A deck of cards. You'll need to perform this one on a table.

Secret

You know one card that is already on top of the deck and you name this card as the first card that you lift and look at. While you look at the first card, you're actually remembering it and will state it as the second card. Magicians call this technique "one ahead" because you're always a step ahead of the spectator. This technique is popular in mentalism.

Skills

The ability to glimpse and recall the bottom card of the deck. For this, a Hindu shuffle is particularly good.

Preparation

None.

Performing the Trick

Ask the spectator to mix up the deck. As you take back the deck, comment that the deck has different cards and that it is indeed mixed up. Continue to cut the cards and then perform a Hindu shuffle or Charlier cut and view and remember the bottom card (for example purposes, we'll say the five of diamonds, as in Figure 5.7).

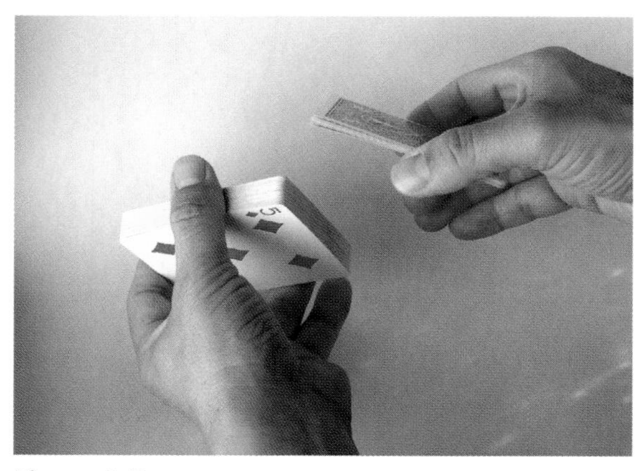

Figure 5.7

View the bottom card, in this case, the five of diamonds. We've used a Hindu glimpse as taught in Chapter 4.

With the bottom card in mind, perform an overhand or Hindu shuffle and bring the bottom card to the top of the deck and set it on the table.

Ask the spectator to cut the cards into two piles. Be sure to track which pile has your known card as the top card.

Announce the card that you remembered (five of diamonds) and then pick up the top card that you don't know. This is shown in Figure 5.8. Look at the top card without letting spectators see it and immediately call out the name of that card (say, "queen of spades") and then pick up the card that you already know (five of diamonds) and look at this card. Again, don't let spectators see this card. You'll want to perform this without hesitation. You're simply naming each card before you look at them in succession.

Figure 5.8
Name the card that you know but briefly look at the card that you don't know. You now know the name of the second card.

Reality Check

You are using the technique of "one-ahead" by knowing ahead of time one of the cards. You've also used a technique known as miscalling, naming something different than what you have at hand.

You've named the five of diamonds but lifted up the unknown card, essentially calling it the "five of diamonds." In the process, you've learned the identity of the unknown card, the queen of spades, and will miscall this card for the second card (five of diamonds) that you look at.

Square the cards together and lay them onto the table. Review how the deck was mixed and cut by the spectator, and the remote odds that one could freely name two specific cards that would end up on the top of each packet.

Turn over and show the two cards to the spectator.

Impossible Predictions and Locations

A POPULAR GENRE OF CARD TRICKS involves having a spectator select a card and then you reveal that you somehow knew all along the card that the spectator picked. Or you make the spectator's card travel to a location that would have been impossible for the card to travel to. In many instances, these tricks rely on a force, causing a spectator to think that he is freely selecting a card when in fact, he is choosing the card that you want him to.

You learned how to force cards in Chapter 4. You're free to use the force that you are most comfortable with and works best for you.

With the following tricks, we'll explain the outcome or revelation and assume that you can force the card.

> **Hint**
>
> I always like to perform a quick card trick before performing a "force" trick. I think that psychologically it shows spectators that you can find their selected cards, which leads to a good forced card outcome.

Simple Prediction

There's no better proof of being able to predict the future than by first writing it down. Here is such an effect.

Effect

You bring out a piece of paper and tell spectators that you've already written the name of a card on it. Or, you can write a prediction on a piece of paper as spectators watch you, but not allow them to see the name of the card that you're writing down. With either approach, fold the paper or turn it face down so spectators can't read it.

Performing the Trick

Force the card that you've written down. Have the card returned to the deck.

> **Hint**
>
> With forced card tricks, I think it's best to wait a bit between the time when you've returned the card to the deck and before you reveal that your written prediction is correct. Just a few suggestions: 1) you can talk about trying to predict things and having the luck of a weatherman, 2) you can try to read their minds to see if you can guess their card and fail, or 3) you can attempt to find their card and fail and then say that you have a backup plan that involves the slip of paper.

Ask the spectator to name their card.

> **Hint**
>
> In a forced card trick, it's always stronger to have the spectator name his card before you reveal it. It adds to the drama and impact.

Allow the spectator to open your prediction and see that you were correct.

You can be ready to perform this trick at any time by carrying around in your wallet, briefcase, purse, or book bag a folded piece of a paper with the name of a card.

Another strong effect is to bring out the paper already sealed within an envelope that is held by a spectator. The sealed envelope reinforces the fact that there is no way that you could have switched the prediction. Also, the sealed envelope prevents a spectator from looking at the prediction before you're prepared to reveal it.

As you'll see in the chapter on stage magic, you'll learn how to perform a forced card trick that works in stand-up situations.

Animazement

This card force trick works with an animation that is a part of this book. If you turn to the back of the book, you'll notice that there's a small drawing that appears on the top-right corner of every page throughout the appendix. And if you use your thumb to flip through the pages, back towards the front, you'll notice that each picture forms part of a short animation that ends in a card prediction.

Simply force the seven of diamonds.

After failing to figure out and identify their card, tell spectators that you're going to have to refer back to the book to see what you did wrong.

Point out the tiny pictures in the top-right corners of the back pages of the book.

Ask the spectator to name his card.

Flip through the pages to display the animation. Show that the animation reveals his card.

Picture Yourself as a Magician—1

©istockphoto.com/Sean Locke

We've taken the title of this book and created a card trick. Here, you act as if you can't recall how to perform a trick and need to refer back to this book. Turn to Appendix A and ask a spectator to read the instructions. You can also watch a performance on the DVD.

The instructions are designed to be funny. In the routine, you follow the directions to force a card (spectators don't know that you are forcing a card), and then it appears as if you get into trouble. As the spectator reads, she mentions props that you'll need, items that you don't have at hand, and there will be steps that you have neglected to do.

At the end, the chosen card is revealed on a page at the end of the routine, which the reader shows to the audience.

Skills

Cross-cut force.

A false shuffle.

Preparation

Place the two of spades on the top of the deck.

Follow the instructions and during the cutting segment, force the two of spades. The instructions will require that you use a cross-cut force.

I recommend that you read through the entire script in appendix A so you can learn what steps are presented. The routine is funniest when you act as if you are in trouble, as when there are props that are missing and steps that you have forgotten.

Appendix A does provide a brief explanation on the first page. When you give the book to the spectator to be read out loud, you want to turn to the next (second) page and have the spectator begin there.

Picture Yourself as a Magician—2

Appendix B provides a second revelation. This is a comedy routine where you use the book to reveal the chosen card and offer a series of funny gags.

For example, one of the gags asks a spectator to think of any four-legged animal. When you tell the spectator that you have the animal printed in the book, you display a picture of a simple stick figure animal that has four legs. While the gags may sound cheesy during an initial read, if you build them up properly as if you've somehow got the answer printed in this book, you'll get lots of laughs. Trust me on this one.

To help with the routine, I've written in small type the picture that will be revealed on the next page. For example, in the picture before the animal, you'll notice in small type at the bottom of the page the statement "name any four-legged animal." Select a spectator to name any four-legged animal. After the spectator names an animal, turn the page to show your "revelation" on the next page.

The second to the last page simply states "card" on the bottom page to alert you to the fact that the next page shows a large picture of the four or hearts. At the end, the chosen card, the four of hearts, is revealed on the last page in Appendix B, which you show to the audience.

Simply force the four of hearts and then open the book to Appendix B and follow the cues and ask spectators a question and then reveal the associated gag.

This routine works because you show the audience some funny gags that momentarily allow them to forget about the card routine. And at the end, you deliver by revealing their card printed in this book.

Lost and Found

Here's a trick that combines a card control and force. You actually make the spectator find his own card that's been lost in the deck.

Effect

A spectator selects a card and returns it to the deck. The spectator is asked to cut the deck to select a second card. When he cuts the deck, he locates his own card.

Materials

A deck of cards.

Secret

You control the card and then force it back to the spectator.

Skills

Double-undercut.

Cross-cut or deeper-cut force (you'll need a table if you're using the cross-cut force).

Preparation

None.

Performing the Trick

Allow a spectator to freely select a card. Have the card returned to the deck and hold a break with your pinky.

Perform a double-undercut to bring the selected card to the top. If you like, you can perform an overhand or Hindu shuffle control.

> **Reality Check**
>
> The spectator's selected card is now on top of the deck and ready to be forced.

Perform either a cross-cut or deeper-cut force. The spectator finds a new card and discovers that it's his first one.

Gambling-Themed Card Tricks

MAGIC TRICKS THAT ARE BASED on gambling are popular. While card moves used in cheating don't necessarily resemble those used in magic, although there is some cross over, with poker being so popular these days, any trick with a poker theme will intrigue those who are into cards.

Here are a couple that you can have fun with.

The Ten-Card Poker Deal

This is a classic. With only ten playing cards, you deal two hands of five-card poker and win three times in a row.

Materials and Requirements

You'll need ten cards from the deck: three three of a kinds and one indifferent card that doesn't work with any of the other cards to help form a hand. For purposes here, we will be using three nines, three eights, three sevens, and one six, as shown in Figure 5.9. You can use cards of any suit.

Figure 5.9
The setup for Ten-Card Poker.

Preparation

Take a pencil and place a dot on each of the white corners of the odd card's back, in this case the six. It's important that you ensure that you can see these dots under your performing conditions. If necessary, you can use an ink pen that matches the color of the back and place a tiny dot in each corner just extending the back design.

Secret

Whichever player receives the indifferent card, the six, loses the game. You have to deal the cards so the spectator is dealt the indifferent card. When the spectator has the indifferent card, you'll always win with a three of a kind versus two pairs, two pairs versus 1 pair, and occasionally, you'll win with a full house.

Performing the Trick

Start out with the six, the indifferent card, on top of the pack of ten cards. The order of the rest of the cards doesn't matter. If you want to mix the cards before the deal, a good way is to use an overhand shuffle. Just mix the cards until you see the indifferent card and then shuffle it on top of the pack. With only ten cards, the indifferent (marked) card is not hard to find.

Deal the cards giving the spectator the top (indifferent) card and then dealing the rest of the cards, as in Figure 5.10. Turn over your cards and show that you have won.

Figure 5.10
The spectator gets the six and loses the round.

As you gather the cards, note where the six (indifferent card) is and make sure that it ends up on top of the pack. Or if you like, you can simply gather the cards and perform an overhand shuffle to control the indifferent card to the top.

Deal the cards and give the spectator the six and you'll win again.

Tip

For the second deal, you can control the indifferent card to the second spot on the deck and have the spectator deal the cards. With the indifferent card in the second position, the spectator will receive the odd card. This is probably stronger although you will have to gauge your spectator to ensure that they won't try to trip you up.

For the third stage, have the spectator mix the cards. Ask the spectator to spread the cards on the table. Look for the location of the indifferent card. If it's in an even location, tell the spectator to deal the cards; he'll still get the indifferent card. If the card is in an odd location, you'll have to gather the cards in order and deal. In either event, you win again.

Hint

Poker-themed tricks where magicians always win can get a bit tiresome. Sometimes, it's fun to pit friends against each other or husbands against wives and render the outcome you want. For example, the wife always wins.

Dealing a Royal Flush

This effect allows you to deal yourself a royal flush under a guise where you explain how gamblers cheat by dealing cards from the bottom of the deck.

Effect

You deal five hands of five-card poker and show how a cheater would pull cards from the bottom of the deck, which become your hand. You show that you have dealt yourself a full house. You gather the cards, perform a cut, and then decide to deal a second hand. After this deal, you've dealt yourself a royal flush.

Materials and Requirements

A deck of cards. You'll need to perform this one on a table.

Secret

The process of exposing the bottom deal actually allows you to arrange the order of the deck so you deal yourself a royal flush.

Skills

False cut (optional).

Preparation

Gather four of the cards necessary for a royal flush, ten through king, in spades, as shown in Figure 5.11. Place these on top of the deck before you begin the trick.

Figure 5.11
Place a partial royal flush on top of the deck.

Performing the Trick

Talk about cheating and openly go through the deck and gather five cards that make up a full house. Make sure that two of these cards are aces and make sure that one of the aces is the ace of spades to complete the royal flush that you've set on top of the deck, as shown in Figure 5.12. (We filled out the full house with sevens in the example.)

Show these cards to your spectators and openly place them on the bottom of the deck.

Figure 5.12
Stack a full house on the bottom of the deck that contains two aces. Make sure that one of the aces is the ace that finishes your royal flush.

Explain how gamblers have a way to deal cards from the bottom of the deck. Deal five hands of five card poker with the fifth hand going to yourself. When you deal to yourself, blatantly pull the card from the bottom of the deck and deal it to yourself.

Show that you've dealt yourself a full house. You need to make sure that the ace of spades is on the bottom of your hand when you turn the cards face down. Gather the other cards and be sure not to change their order. Stack the other hands on top of your cards on top of the deck.

Reality Check

You have the cards that make up a royal flush in every fifth position in the top 25 cards. When you deal five hands again, the fifth hand, your hand, will be a royal flush.

Talk about how a real gambler would be able to perform the dirty work undetected. Perform a false cut and deal five hands of five card poker again.

Show that you've dealt yourself a royal flush.

More Card Tricks

We provide some more card tricks in later chapters. Look for the "Missing Pip" trick in Chapter 9, which covers stage magic. There's an amazing trick in Chapter 10 that covers street magic, called "KISS," where a female spectator blows a kiss and it lands on her selected card. Another strong trick in Chapter 10 is "The Billfold," where a spectator's card ends up in your wallet. These card tricks will make use of your sleight-of-hand and card handling skills.

Next Steps

In the next chapter, we discuss popular trick decks. Even though you already know how to perform magic using a real deck and applying some slick sleight-of-hand, you may enjoy working with a trick deck.

Trick Decks

YOU'VE ALREADY LEARNED HOW to perform tricks with a real deck of cards, and you can execute basic card sleight-of-hand. In this chapter, we provide an education in the use of gimmicked or trick decks. You'll be amazed at the additional things you can do.

A gimmicked or trick deck is one that is not normal. Like a modified, custom car, the deck is literally "tricked out."

Audiences are familiar with the concept of a trick deck. Many have probably seen demonstrations of such decks and may have even purchased them to play around with. While knowledge of such decks can pose a problem, if you present a trick correctly, a spectator in the know will probably never consider a trick deck.

As a pro, I prefer to use a real deck because I can freely use it under any circumstance and hand cards out for examination at any time. My hands do all the work.

But under certain circumstances, I have relied on a trick deck. Recently, when filming a short magic segment for a local television show, I used a trick deck to accomplish my minor miracle. There's absolutely nothing wrong with using trick decks. But you have to use them in the right manner, which we'll explain here. And you also have to practice using them just as with any magic trick.

Using a Trick Deck in the Correct Manner

MANY BEGINNING MAGICIANS or those who are dabbling in magic tend to blatantly use a trick deck until there's no other explanation. Even more suspect is the beginner who brings out a deck of cards that has a case that says "magic cards."

I recommend that you never use a trick deck as your only trick. It places too much heat on the deck and will cause spectators to want to examine the cards. It's always best to perform magic in sets, a series of tricks so you can take the trick deck out of play and move onto something else.

Another suggestion is to make sure that your trick deck and real deck look exactly alike, preferably with a readily recognized design, for example, red or blue Bicycle backs. (All of the decks that we discuss here come in Bicycle and other popular versions.) The idea is to mix up your sets so that you first perform a trick using a real deck that allows a spectator to freely shuffle the cards, and in the process, assures them that the deck is normal without stating so.

After performing a trick with the real deck, you place it into your pocket or case and then perform another trick, say a coin trick. After the coin trick, you seemingly pull out the same deck to perform another card trick, but actually bring out the trick deck. You switch decks.

This disarming technique will convince spectators that you are using the same deck when you've introduced a trick one. You're ready to do some magic.

In addition, where applicable, we'll teach you a few moves that will disguise the use of a trick deck.

One-Way Force Decks

PERHAPS THE BEST KNOWN of the trick decks is the one-way force deck. The concept is simple. This deck is simply one that consists of 52 of the same cards, for example, the king of hearts. You can see a picture of a one-way force deck in Figure 6.1. The spectator can freely choose a card as you spread them face down and will always select the force card.

Figure 6.1
A one-way force deck features a deck with all of the cards the same.

Use a one-way force deck as an easy way to force a card on a spectator. The deck can support the force tricks explained in Chapter 5.

Lay audiences may suspect the use of such a deck, but if you bring a one-way deck out only momentarily, and use it to follow a trick that uses a real deck that has the same back, you should encounter few problems.

The great thing about a one-way force deck is that you can ask a spectator to select a card and return it to the deck. After you put the deck away, you can go about making the grand revelation of your choice, as explained in the force tricks in Chapter 5.

Hint

One way to mask the use of a one-way force deck is to place an indifferent card on the bottom of the deck that the spectator inadvertently sees. The card on the bottom should contrast with the force card. So, for example, if your force card is the king of hearts, you'll want a black or red number card on the bottom, as shown in Figure 6.2.

Figure 6.2
Disguise a one-way force deck with the use of a different card that the spectator sees on the bottom.

The odd card on the bottom reinforces in the minds of spectators that the deck contains different cards. When you present the cards so a spectator can select one, just make sure that they don't select the bottom card. And should a spectator insist on selecting the bottom card, you already know it and can use this information in an impromptu trick.

You can quickly read their mind and name the card that they selected (the bottom card), and then have them select another card where you won't cheat. Many times, in magic, you do have to be flexible and readily change directions to suit the situation.

The Svengali Deck

THE BEST KNOWN OF THE mechanical trick decks is the Svengali deck. This deck is often demonstrated at fairs and sold by the thousands to the general public. In fact, if a spectator knows about trick decks, this is the likely one that they will suspect.

The Svengali deck is one that features 26 force cards that are alternated between 26 regular cards. Actually, most Svengali decks on the market contain 48 cards with 24 like force cards and 24 real cards. Figure 6.3 shows a Svengali deck.

Figure 6.3
A Svengali deck features a force card as every other card. In this particular Svengali deck, the force card is the king of hearts.

In addition to the setup of alternating cards, the force cards are physically cut a bit shorter than the other cards, which gives the deck some unusual properties, as shown in Figure 6.4.

Figure 6.4
In a Svengali deck, the force card, in this case the king of hearts, is slightly shorter than the other cards.

There are a series of tricks that can be performed only with a Svengali deck, but these are tricks that many lay people may recognize. We're going to discuss how you can use a Svengali deck to perform great magic.

Svengali Properties

BECAUSE EVERY OTHER CARD in a Svengali deck is a real one and the cards have been cut to alternate between long and short, you can thumb through the cards and seemingly show that every card is different. Generally known as "running through the deck," the move is unique to working with a Svengali deck and is similar to riffling through the cards in a riffle shuffle.

By running the cards in one direction, you can seemingly show the cards to be different. But when a spectator selects one of the cards running by, it will always be the force card. You see, because of the shorter force cards, the cards fall in pairs with the real card on the bottom and the force card on top.

So anytime you or the spectator cuts the deck, assuming that you cut the cards by grabbing the short ends of the cards, you will always cut to the force card. As you'll see, we'll make good use of these properties in tricks.

By turning the deck face-up and running through the cards, you can make it look as if all of the cards have turned into the force cards. While this is quite impressive, I think it goes too far. Those in the know about such trick decks will immediately identify your deck as a Svengali deck that they've seen or purchased at the fair. And even to those who don't know about such trick decks, they'll immediately realize that there's something "different" about the cards.

One important aspect: when you have a chosen card returned to the deck, it's important that it be returned to the exact spot where it came from. This ensures that the alternating order of the cards isn't changed. Simply hold the place in the deck where the spectator selected the card and have it returned to the same place.

Hint

When using a Svengali deck, it's best to run through the cards (explained in the next section), ask a spectator to say "stop," and then simply slide the card out for everyone to see and remember. You hold the selected card and display it and place it back into the deck. The less that spectators handle the Svengali cards, the better.

If you allow a spectator to remove a card and show it, they may refuse to insert the card back into the deck at its designated place. As a result, simply ask a spectator to say "stop" as you run through the cards, and then slowly and clearly display the card at that location. And before you display the card, you can verify that the spectator is happy with his selection. And if he isn't, simply run through the cards again or continue running from your current location.

Svengali: The Moves

HERE ARE THE BASIC MOVES that you'll need to work with a Svengali deck.

Running Through the Deck and Forcing a Card

This move serves two purposes: 1) you show that the cards are different, and 2) a spectator can freely choose a card but will select a force card.

This move is much like riffling the deck, as explained in Chapter 3, but unlike when you shuffle, you are riffling the entire deck in one hand. Also, you are bowing the cards less than you would in a riffle shuffle and letting go of the cards so they fall into your other hand. As mentioned earlier, the varying lengths of the cards allow the cards to fall in pairs with an indifferent card on the bottom and the force card on top, as shown in Figure 6.5.

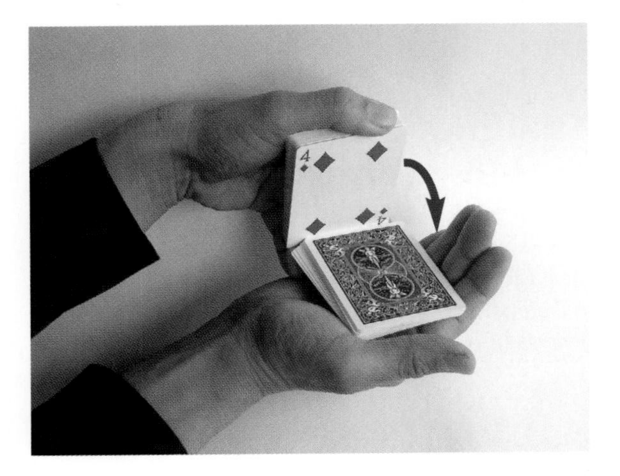

Figure 6.5

When running the cards, the cards fall in pairs with the indifferent card on the bottom and the force card on the top.

When a spectator says "stop," the top card at that location will be a force card. At this point just stop your riffling and push out the card and display it to spectators, as shown in Figure 6.6.

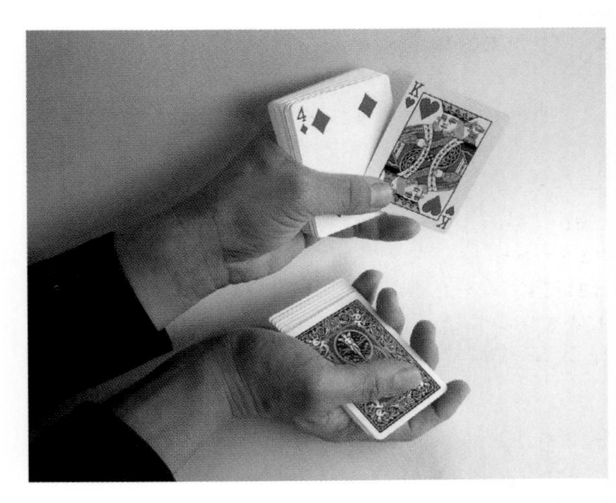

Figure 6.6

After a spectator says "stop," pull out and display the card to spectators.

Hint

Something that ensures that a spectator has selected a force card, particularly when using some of the later methods, is to subtly mark the cards so you can immediately tell which ones are the force cards.

The easiest way to do this is to take a pencil and place a light dot in the upper-right and lower-left corners on the back of each force card in the white border area of the card. You'll want to make sure that the dot is visible under the lighting in your performance venue. You can always look for this pencil dot and rest assured that the spectator has selected a force card. Figure 6.7 shows how to mark a card.

Figure 6.7
Mark your cards with a light pencil dot in the corner of each force card.

Cutting the Cards

If you cut the cards by holding the short ends of the deck, you will always cut a force card to the top of the deck, as shown in Figure 6.8. Better yet, when you cut a Svengali deck in this manner you don't change the order of the deck for your purposes. Cut the deck and you've given the deck a legitimate cut, but the deck is still in order to do tricks.

Figure 6.8
If you or the spectator cuts the deck by holding the short ends of the deck, it will cut the deck at a force card.

Shuffling the Cards

The Riffle Shuffle

Another thing you can do with a Svengali deck is riffle shuffle the cards. Again, as when cutting the deck, you do mix the deck, but because the cards fall in pairs, the deck is still in order. As long as the force and real cards alternate, the deck is ready to go.

One important thing for the riffle shuffle: you have to make sure that you cut the deck with a short card on top of each pile. For this reason, the second variation of the riffle shuffle that's taught in Chapter 3 where you riffle cards into your hand to split the deck has a definite advantage.

Figure 6.9 shows how to split the deck when riffle shuffling a Svengali deck.

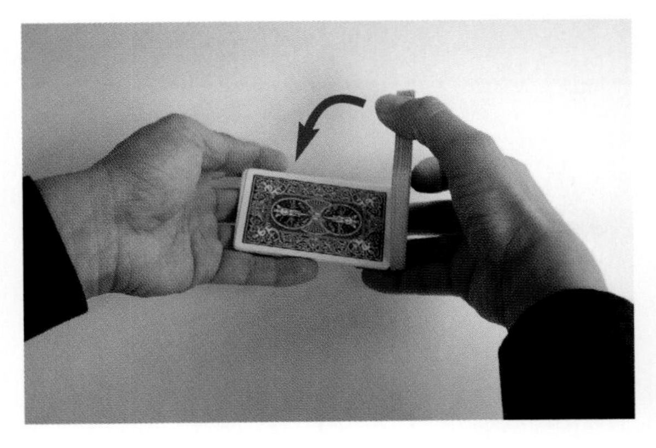

Figure 6.9
You can riffle shuffle a Svengali deck and keep it in its necessary order. It's best if you split the deck by riffling the cards, a method that's taught in Chapter 3.

You'll want to practice shuffling a Svengali deck until you're comfortable that the cards are falling in order. You'll probably find that if you shuffle more slowly, you're less likely to get the cards out of order.

The Overhand Shuffle

Another shuffle that works with a Svengali deck is an overhand shuffle, although you have to alter your technique somewhat. In the overhand shuffle, which is taught in Chapter 3, you hold the deck with your right hand with the fingers on the back edge of the deck and the thumb on the front edge.

As in a conventional overhand shuffle, you release packets of cards from your right hand into your left, but when using a Svengali deck, instead of simply releasing the cards, you perform something akin to a small riffle by slightly squeezing and bowing the cards to ensure that the cards fall in pairs. I like to use my right thumb to do this, as shown in Figure 6.10.

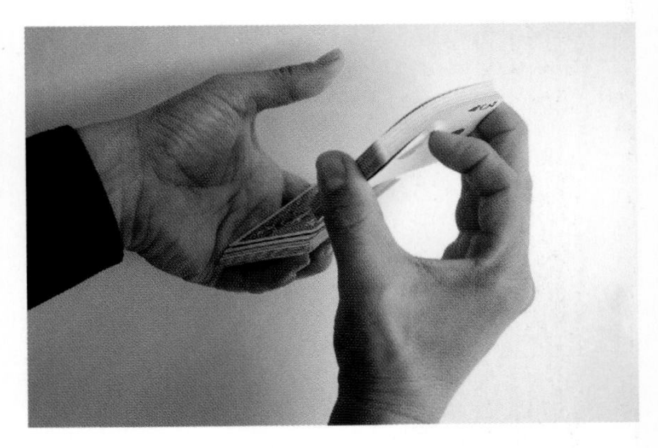

Figure 6.10
Squeeze and release the cards from the right hand when performing an overhand shuffle with a Svengali deck. Notice how the deck is slightly bowed in the right hand. This ensures that the cards separate in pairs.

There is a definite knack to this technique that will require practice. Also, you have to make sure that the cards land in your left hand in pairs. But if you can master this one, you'll be able to perform a fair overhand shuffle and still maintain the necessary order of the Svengali deck.

A Second Force

AFTER RUNNING THROUGH the cards to show that they're different, you can ask a spectator to cut the cards. But before you do this, you have to make sure that the spectator will grab the short ends of the deck, as explained earlier.

Hint

When you're performing an earlier trick where the order of the deck doesn't matter, ask a spectator to mix the cards, place the deck on the table, and then cut the cards. If the spectator already cuts the deck by its short ends, he'll probably cut the Svengali deck the same way to reveal a force card.

Another thing you can do is tell the spectator to cut the Svengali deck and then subtly show him how to grab the deck by the short ends. Don't actually tell him to grab the deck by the short ends, just show him how to do it and most will follow your lead.

Don't worry if the spectator doesn't follow your lead and cuts the deck the wrong way by holding the long sides of the deck. There's still a fifty percent chance that he cut to a force card.

You can look at the top card to verify that it is a force card. And if it isn't, no problem. Just act as if you were asking the spectator to cut the deck to mix it and then run through the deck as mentioned in the earlier section and ask a spectator to touch a card. You've given yourself two opportunities to force the same card with the Svengali deck.

A Third Force

IN THIS FORCE, you deal cards onto the table until a spectator says "stop." And before the spectator turns over her card, you can show that the card immediately before and after her selected card is different.

Hint

You'll have more confidence with this force if you mark the force cards as explained earlier.

Take your Svengali deck and riffle or overhand shuffle it and then run through the cards to show that the cards are different. If you like, you can cut the cards, or if you've already verified that the spectator will cut the cards by grabbing the short ends, have the spectator cut the deck.

Begin dealing cards from the top of the deck onto the table. Deal a few cards and then tell the spectator to say "stop" at any time. Most spectators will say stop after a few cards. Some may go ten or more cards. Just as long as you don't run out of cards, you can continue dealing until the spectator says stop.

If you get more than halfway through the deck, you may want to put the deck back together and start over, and ask the spectator to say "stop," sometime today.

When the spectator says "stop," secretly note where the closest force card lies (you can tell from your marks if you have them). The force card will either be in the pile that you're dealing from in your hand, or on top of the pile on the table, as shown in Figure 6.11.

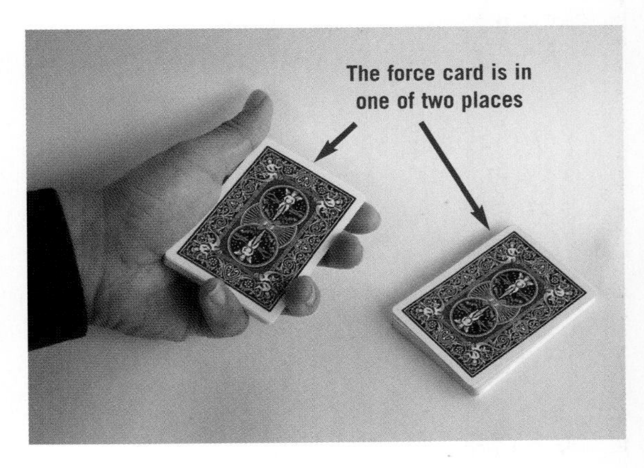

Figure 6.11
The force card will either be in the pile that you're dealing from in your hands or on top of the pile on the table. If you have them, you'll want to look for your marks that designate your force cards.

Take the force card and without showing it, set it aside face down on the table.

Note that you don't give the spectator a choice of which card they want. You simply take the card that you want without saying a word and set it aside on the table.

After setting the force card onto the table, show that the card immediately before was an indifferent card, as well as the card immediately after it. Because of the way a Svengali deck is structured, the cards before and after will be different from your force card. You can say, "if you said 'stop' one card earlier, you would have chosen this card, and if you had said 'stop' one card later, you would have chosen this card." This is shown in Figure 6.12.

Figure 6.12
You can show a spectator that the cards immediately before and after his selected card are different.

Gather the cards and place them back into the box. You want to move the Svengali deck out of play. Have the spectator turn over the force card so everyone can see. Go into your favorite force revelation, from those that were explained in Chapter 5.

Svengali Revelations

Spectator Finds His Own Card

You can simply combine Svengali forces so a spectator chooses the force card and then freely selects another card, and effectively finds his own card. As an example, you can run the cards to allow a spectator to select a force card. And then use the third force where you deal cards, to allow a spectator to "find" his card again.

Materials and Requirements

A Svengali deck.

Skills

The ability to force a card in two different ways using the Svengali deck.

Shuffling a Svengali deck (optional).

Preparation

This one can be done at any time and there is no preparation to the deck.

Performing the Trick

Use one force, for example, running through the deck to allow a spectator to choose the force card.

Mix the deck using the riffle shuffle (optional) and cut the cards.

Use a second force, for example, dealing through the cards, to allow the spectator to select his own card.

This is an adequate trick, but there's so much more that you can do. Here are some ideas.

The Prediction

Effect

You bring out an envelope and mention that there's a prediction inside. The envelope is left in plain view or given to a spectator to hold. A spectator selects a card. When the envelope is opened, a note inside indicates the chosen card.

Materials and Requirements

A Svengali deck.

An envelope and paper.

Skills

The ability to force a card using the Svengali deck.

Riffle shuffling a Svengali deck (optional).

Preparation

Write down the name of the Svengali force card on a piece of paper and place this inside of the envelope, as shown in Figure 6.13. You can seal the envelope if you wish.

Figure 6.13
Your prediction, which is the force card, goes into the envelope.

Performing the Trick

Bring out the envelope and set it onto the table and leave it in plain view. If you like, you can give the envelope to a spectator for safe keeping. Shuffle the Svengali deck if you like and then use it to force the card by your favorite technique. Display the chosen card to the spectators, place it back into the deck, and put away the deck.

Ask the spectator to open the envelope and read the prediction.

Card to Impossible Location

A Svengali deck allows you to easily force a card and then reveal that it has traveled to an impossible location. And the advantage of using a Svengali deck is that you can run through the cards to show a spectator that their card has vanished from the deck. It's a pretty cool trick. We'll explain a selected card that vanishes to a card box.

Effect

A spectator's selected card vanishes from the deck and is found inside of the card box.

Materials and Requirements

A Svengali deck.

An extra card that matches the force card of your Svengali deck.

Skills

The ability to force a card using the Svengali deck.

Shuffling a Svengali deck (optional).

Preparation

Place the extra matching card on the face of the Svengali deck and place the entire deck into its box.

Performing the Trick

Remove the Svengali deck from its box and leave behind the matching (extra) card. Leave the card box closed on the table in plain view.

Riffle or overhand shuffle the Svengali deck if you like and then use it to force the card by using your favorite technique. Display the chosen card to the spectators. Snap your fingers and then say that the spectator's card has left the deck. Run slowly through the cards to show that the spectator's card is no longer in the deck.

Ask the spectator to pick up and open the card box to find that his selected card has left the deck and is now in the card box, as shown in Figure 6.14.

Figure 6.14
The spectator's selected card is now found inside the card box.

Card Stab

You can make a spectator select their own card by having them insert a pencil, pen, finger, and more into the deck as you run through it. But a classic of magic is covering the deck with paper and allowing a spectator to gently stab through the side edges with a letter opener (or knife, if you're so inclined) and show him that he found his own card. Here's how to perform this classic using a Svengali deck.

Effect

A spectator selects his own card by stabbing the deck that's been wrapped with paper.

Materials and Requirements

A Svengali deck.

Paper.

A letter opener.

Skills

The ability to force a card using the Svengali deck.

Shuffling a Svengali deck (optional).

Preparation

None.

Performing the Trick

Riffle or overhand shuffle the Svengali deck if you like and then use it to force the designated card using your favorite technique. Display the chosen card to the spectators and place it back into the deck. Mix up the deck by shuffling and then cutting it.

Wrap the deck in a sheet of paper, as shown in Figure 6.15. If you like, you can also use a napkin or paper towel, which will be easier to tear away later.

Figure 6.15
Wrap the deck in a sheet of paper.

Ask the spectator to insert a letter opener through the paper into the deck, on the short end of the deck, as shown in Figure 6.16.

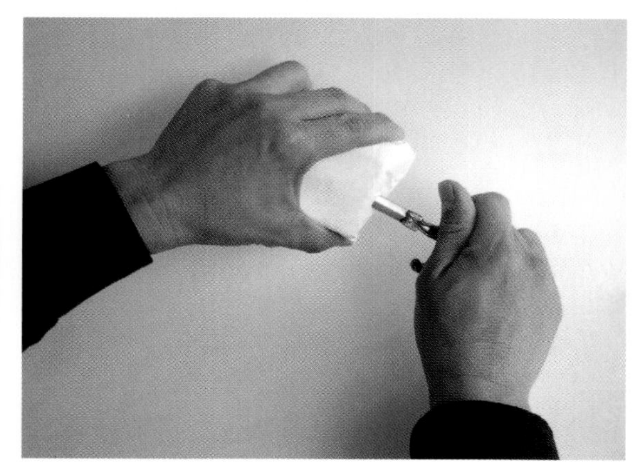

Figure 6.16
The spectator inserts the letter opener through the paper into the deck.

> **Reality Check**
>
> Wherever the spectator inserts the letter opener, the force card will either be above or below the letter opener.

Unwrap the paper maintaining the position of the letter opener. You will probably have to rip away the paper, as shown in Figure 6.17.

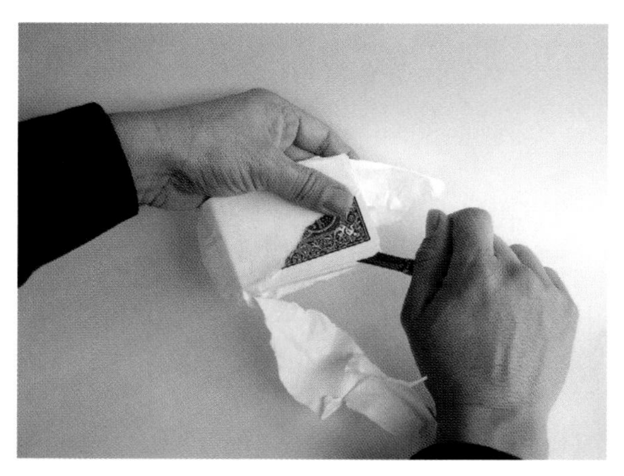

Figure 6.17
Rip away the paper from the deck while maintaining the location of the letter opener.

As you lift the top half of the deck above the letter opener, look for your marking on the card below the letter opener. If it's one of the force cards, set the top half down on the table and then ask the spectator to turn over the card immediately under the letter opener, as shown in Figure 6.18.

Figure 6.18
If the force card is underneath the letter opener, remove the top half of the deck and allow the spectator to turn over the top card to show that it is his card.

If you glimpse at the top card of the bottom half and determine that it's not a force card, the force card is the bottom card of the top half, as shown in Figure 6.19.

Figure 6.19
If the force card is not underneath the letter opener, it's the bottom card of the packet that you're holding.

In this instance, simply lift the top half of the deck above the letter opener and show that the bottom card is the spectator's card. The spectator has somehow inserted the letter opener into the exact location where his card resided.

Card at Any Number

This is a somewhat esoteric trick because after a card is selected, the spectator freely names a number between five and forty and then you count through the deck to reveal the spectator's card at that number. The process is actually similar to the dealing force that we taught earlier.

Effect

A spectator names any number and his chosen card is found at that location in the deck.

Materials and Requirements

A Svengali deck.

Skills

The ability to force a card using the Svengali deck.

Shuffling a Svengali deck (optional).

Preparation

None.

Performing the Trick

Have the force card selected.

Hint

Because you'll essentially be performing the dealing force to reveal the spectator's card, it's best to have the spectator cut the deck or select a card as you run through the deck. This way, you're not repeating the moves.

Return the card to the deck and then mix the deck by shuffling and cutting it.

Ask the spectator to name any number between five and forty. There are two scenarios here.

If the spectator names an odd number, the force card will be at that location. Simply count cards until you reach the chosen number and turn it over to reveal the spectator's card.

If the spectator names an even number, the force will be the next card after counting to the spectator's number. Count the cards and then turn over the top card in the pile of cards that remain in your hand to show that it's the spectator's card.

If you've marked your cards, you won't have to count the cards.

Using Dice

"Card At Any Number" also works with dice. You can have a spectator roll anywhere between one and six dice and add up the numbers that are displayed. By using the techniques explained here, you can show that their card is at that location.

Spelling Trick

One more variation, you can spell virtually anything, a spectator's name, their chosen card, their hometown, and more. The technique is similar to that explained for "Card At Any Number."

Ask the spectator to name his card, which will be the force card. Spell the name of the card and then deal cards onto the table, one for each letter. When you get to the last letter of the card, look down and determine where the force card lies. It will either be on top of the deck in your hand or on the table.

If it's in your hand, spell the last letter and lift the top card in your hand and turn it over to show that it is the spectator's card. If the force card is already on the pile on the table, you'll spell the last letter and end up with an indifferent, non-force card. Simply deal this card to the pile on the table and then turn over the next card in your hand, which will be a force card. If you've marked your cards, this will help.

Stripper Deck

ANOTHER POPULAR TRICK DECK is what's known as the stripper deck. This deck features cards that are not true rectangles, but are actually trapezoids—one side is shorter than the other, as shown in Figure 6.20.

When the cards are facing the same direction and completely in line, you can quickly locate a single card that's facing the other way by feeling for the protruding edge, as shown in Figure 6.21. As you can imagine, this is useful to magicians.

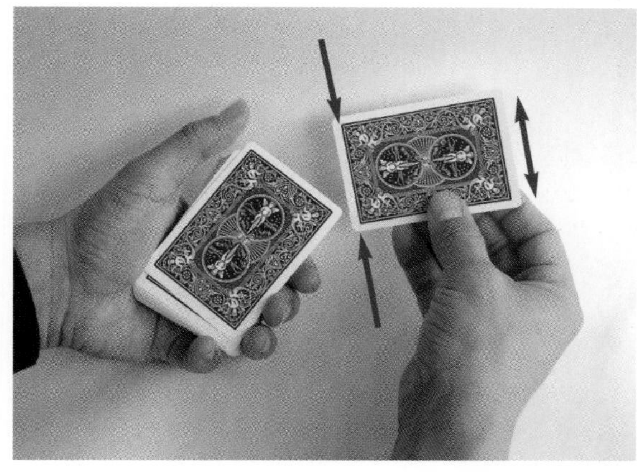

Figure 6.20

In a stripper deck, one edge of each card is shorter than the other.

Figure 6.21

When a card is facing the other direction, you can feel the edge protruding.

The properties of a stripper deck allow you to accomplish the following:

- **You can have a selected card returned to the deck in the wrong direction and be able to find it.**
- **You can overhand and riffle shuffle a stripper deck and still find a card or set of cards that is inserted in the opposite direction. It is this seemingly free handling that is the power of a stripper deck.**
- **If a spectator mixes the cards using an overhand shuffle that doesn't turn the cards, he can freely shuffle the cards and you'll be able to find a selected card that appears to be lost in the deck.**
- **You can mix cards, say red and black cards, and by mixing them in the opposite directions, typically with a riffle shuffle, you can later un-mix them by grabbing the protruding edges. And this applies to suits of cards as well.**

Stripper Deck: The Moves

Shuffling a Stripper Deck

You can freely shuffle a stripper deck using a regular overhand shuffle. Simply shuffle the cards in the usual manner as taught in Chapter 3. And if you determine that a spectator mixes cards using an overhand shuffle, you can simply hand the cards to him and he won't alter the orientation of the cards or lose a selected card or set of cards that are facing the opposite direction.

The Riffle Shuffle

You can freely riffle shuffle a stripper deck, but you have to make sure that you don't change the direction of the cards. For this reason, the second method of performing a riffle shuffle, the flashier version where you riffle the cards from one hand to the other to split the pack, is the riffle shuffle of choice. This is shown in Figure 6.22.

Figure 6.22

If you split the deck by riffling the cards, the riffle shuffle won't change the direction of the cards.

The other riffle shuffle where you split the deck using your fingers and thumbs will change the orientation of the packs and create a mess. This is shown in Figure 6.23.

Figure 6.23

Splitting the deck in this manner to perform a riffle shuffle will change the direction of the cards and won't work for a stripper deck.

Controlling a Card Using a Stripper Deck

When using a stripper deck, it's best to ask a spectator to say "stop" and then slide the card out for everyone to see and remember. The less that spectators handle the cards, the better. This will also make it easy for you to perform the "turnaround."

The Turnaround

While this move sounds like something a financial takeover manager may do, it's simply a method for removing a selected card and then subtly turning the card around so that when it's returned to the deck, it's inserted the "wrong" way in the stripper deck so you can later find it.

Spread through the deck in the normal manner and ask a spectator to say "stop."

Close up the deck and slowly bring out the selected card. As you bring out the card, turn it with your hand, so it's now facing the opposite direction, as shown in Figure 6.24. Show the card to spectators.

Figure 6.24
Subtly turn the card as you bring it out to show spectators.

Ask the spectators to remember the card and then return the card into the deck. The card will be oriented in the opposite direction.

If you allow a spectator to remove a card and show it, they may make it difficult to turn the card or deck around in the correct orientation. As a result, simply have a spectator say "stop" and then slowly and clearly display the card at that location and perform the turnaround move. And before you display the card, you can verify that the spectator is content with his choice. And if he isn't, simply spread through the cards again or continue running from your current location.

Cutting a Stripper Deck

You can freely cut a stripper deck, just as long as you don't change the orientation of the packets. Straight forward cutting maintains the direction of the cards and will also allow you to later find a card or set of cards that are facing the other direction.

Cutting to a Spectator's Chosen Card

Beginners often reveal a selected card that's facing the opposite direction by simply running their fingers along the edges of a stripper deck and just pulling the selected card out, which is a telltale sign of a stripper deck. And just based on this "move" alone, a spectator may figure out how the trick deck works.

For this reason, I think that it's important to learn how to cut to the selected card that's facing the opposite direction to the top or bottom of the deck. The action makes it look as if you are simply cutting the deck, but you're actually feeling the deck along the edges to find the protruding card and then cutting it to the top or bottom. Cutting to the bottom of the deck will feel more natural, but cutting to the top is probably more useful.

Cutting a Spectator's Card to the Bottom

Here's how to cut a selected card to the bottom of the deck.

Have a spectator select a card and return it to the deck. Perform the turnaround so that the spectator's card goes into the deck in the opposite direction. After performing an overhand or riffle shuffle, feel the narrow edge of the deck until you find the protruding card, the spectator's selected card, as in Figure 6.25.

Figure 6.25
Feel for the spectator's card.

Using your thumb to hold one edge of the protruding card and your first finger on the opposite edge, lift the entire deck along with this card and place this half onto the table. This is shown in Figure 6.26. Complete the cut and you now have brought the spectator's selected card to the bottom of the deck.

Figure 6.26
Lift the protruding card and the rest of the deck on top. When you complete the cut, this brings the selected card to the bottom.

From here, you can perform an overhand or Hindu shuffle to bring the selected card to the top of the deck, where you can perform any card trick that requires the selected card be brought to the top. With this ability to cut a spectator's card, you can effectively use a stripper deck in place of a normal deck in any trick that requires you to control a card to either the top or bottom of the deck.

Cutting a Spectator's Card to the Top

Here's how to cut a selected card to the top of the deck.

Have a spectator select a card and return it to the deck. Perform the turnaround so the spectator's card goes into the deck in the opposite direction. After performing an overhand or riffle shuffle (see above), feel the narrow edge of the deck until you find the protruding card, the spectator's selected card, as in Figure 6.27.

Figure 6.27
Feel for the spectator's card.

Using your thumb to hold one edge of the protruding card and your first finger on the opposite edge, push down on the selected card to make a gap above it and lift the entire deck that's above this card and place this onto the table. This is shown in Figure 6.28.

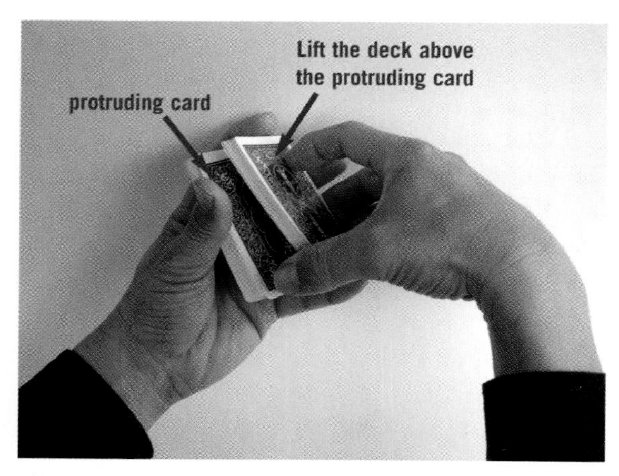

Figure 6.28
Lift the deck above the protruding card to cut the deck.

Complete the cut and you now have brought the spectator's selected card to the top of the deck.

Cutting to Aces (Stripper Deck)

Here's a convincing trick where you cut directly to the aces. Because of the capabilities of the stripper deck, you can freely mix the deck, and under controlled circumstances mentioned above, allow a spectator to shuffle the cards as well. Despite the mixing you find the aces.

Materials and Requirements

A stripper deck.

Skills

The ability to cut to a card that's facing the wrong direction and bring it to the top of the deck.

Preparation

Remove the four aces, turn them around so that they're facing the wrong way, and lay them onto the table.

Performing the Trick

Openly place the four aces into the deck (the wrong way) and mix the cards, as shown in Figure 6.29.

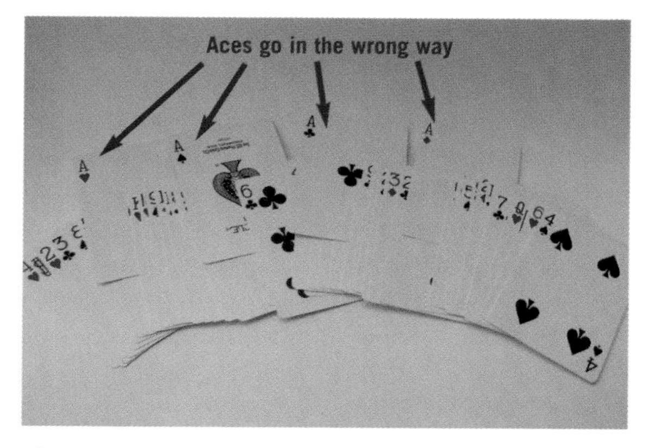

Figure 6.29
Insert the aces the wrong way into the deck.

Use the mixing techniques explained in this section to shuffle the stripper deck without changing its orientation and then cut the cards.

If you've verified that a spectator naturally shuffles the cards using an overhand shuffle, allow the spectator to do so. Also, you can allow the spectator to cut the cards as many times as he wishes.

Check to make sure that one of the aces isn't already on top. If an ace is on top of the deck, which will happen once in awhile, count your lucky stars. You'll be able to use this to your advantage. Simply remove this top card and lay it on the table face down to later show that the spectator somehow shuffled an ace to the top.

Using the technique mentioned earlier, cut the first, topmost ace in the deck to the top, but instead of resting the upper, removed half on the table, place this section under the deck.

From here, you can continue in the same manner, whether an ace was already on top of the deck or if you had to cut an ace to the top.

Cut the first, topmost ace to the top of the deck and rest the section of the deck that you remove onto the table, as shown in Figure 6.30.

> **Reality Check**
>
> The first section that you have placed onto the table has the first, topmost ace on top. The deck now has a second ace on top.

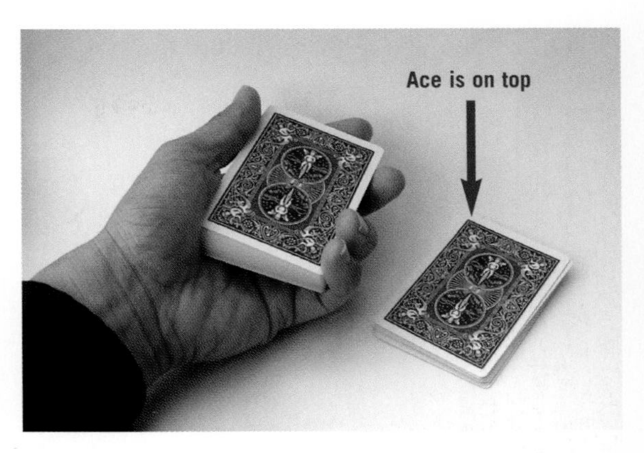

Figure 6.30
Cut the topmost section down to the first ace and place the section down onto the table.

Repeat the process until you have four piles on the table, as shown in Figure 6.31.

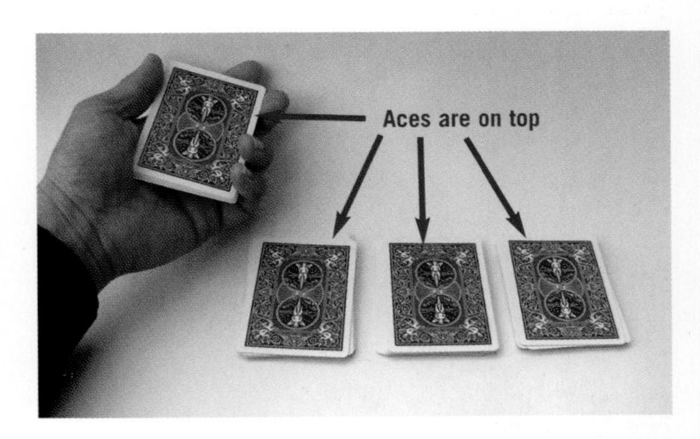

Figure 6.31
Cut the deck four times until you have four piles on the table.

Turn over the top card in each pile to show that they are aces.

Next Steps

In the next chapter, we get into close-up magic that uses rubber bands, coins, and more. These will be great tricks to mix with your card tricks.

7

Close-Up Magic

IN THIS CHAPTER, you'll delve into the world of close-up magic with tricks that use common household items such as rubber bands, coins, and handkerchiefs. With these tricks, you can begin to build an act that combines the close-up card magic that you learned in earlier chapters.

Close-up magic is a type of magic that you perform for small groups of people and usually employs common objects such as currency, coins, finger rings, paper clips, string, and more. Close-up card magic is a subcategory of close-up magic.

Visualize the Tricks

Before you move onto learning and performing the various close-up tricks, you may find it helpful to view the techniques and tricks in action on the accompanying DVD.

For this, please refer to the "Close-Up Magic" video that demonstrates each trick. After watching the video, you may discover that there are tricks that you want to learn first and can move directly to the appropriate section.

Rubber Band Magic

IN THIS SECTION, WE TEACH you an entire rubber band routine that you can learn and perform and consists of several tricks strung together. The routine begins with a trick known as the jumping rubber band, where a band mysteriously jumps between your fingers. And we offer two variations. This is followed by an effect where a rubber band appears to pass through your thumb. And finally, you appear to break and then restore a rubber band.

The Jumping Rubber Band—Version 1

In this routine, rubber bands seemingly jump between the fingers of your hands. We teach you three variations that appear to grow in complexity.

The Basic Jumping Rubber Band

Here's the basic version. You'll find that this trick is fairly well known by the general public, but don't be dismayed. Even though some spectators may know this trick, the second two variations are not as well known and look impossible. The three phases evolve so that each variation looks more difficult.

Effect

A rubber band jumps between the fingers of your hand.

Materials and Requirements

A rubber band. (You'll want a fairly small rubber band and want to test a few to see which size works best for you. I find that a size 16 works well.)

Secret

This effect takes advantage of the elastic nature of a rubber band and the setup causes the band to appear to instantaneously jump between your fingers. In reality, you have set up the band to switch places when you change your finger position.

Preparation

This one can be done at any time and there is no preparation.

Performing the Trick

Wrap the rubber band around the ring and pinky fingers of your left or right hand, as shown in Figure 7.1.

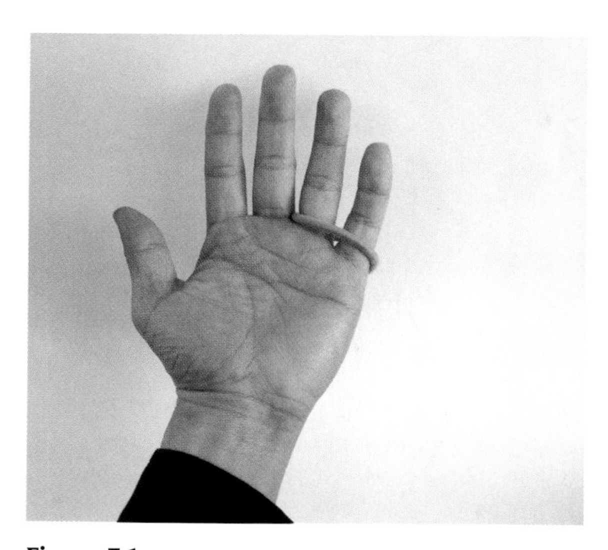

Figure 7.1

Wrap the rubber band around the ring and pinky fingers of your left or right hand. Note that for demonstration purposes and clarity, we are using a hair band in these photos.

With the back of your hand facing spectators, use your other hand to pull the rubber band towards your thumb, as shown in Figure 7.2.

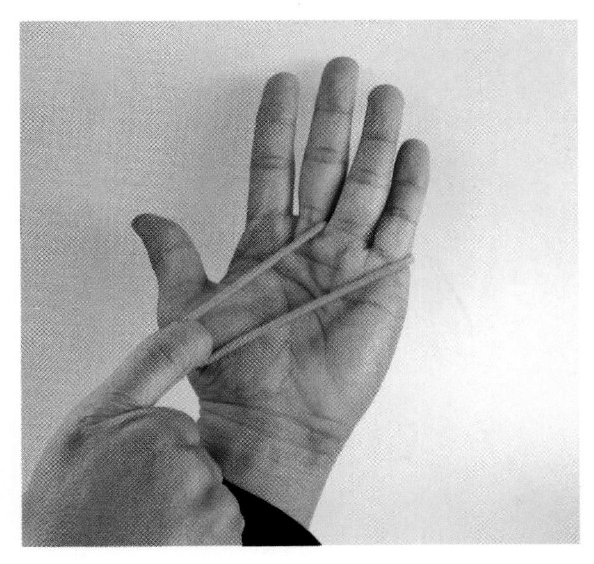

Figure 7.2
Pull the rubber band towards your thumb.

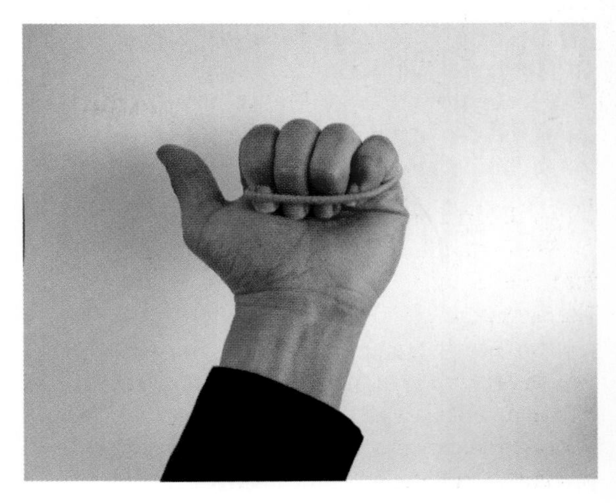

Figure 7.3
Insert all of your fingers into the extended rubber band. Rest the band on the tips of your fingers.

Curl your fingers so your hand begins to form a fist and insert all of your fingers into the extended rubber band. Rest the band on the tips of your fingers, as shown in Figure 7.3. The audience should not see the rubber band resting on your fingers. If you have spectators to your sides, bring your hand in closer to your body to cover this.

By simply opening your hand, the rubber band will jump from your ring finger and pinky fingers to your first and second fingers, as shown in Figure 7.4.

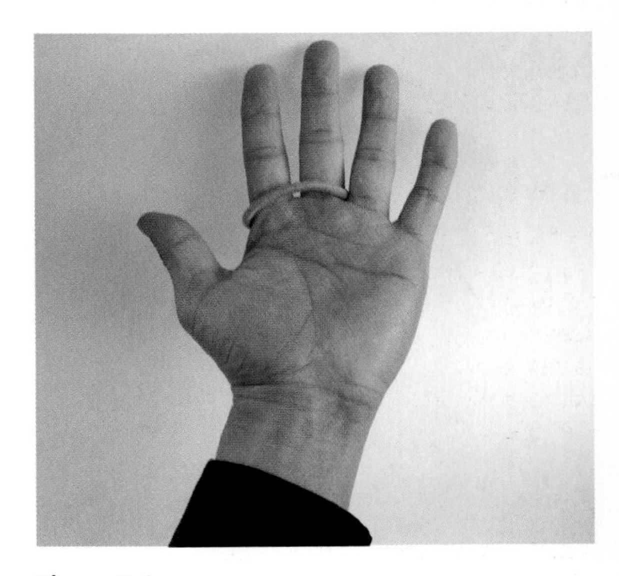

Figure 7.4
Open your hand to cause the rubber band to jump between your fingers.

This is the basic trick.

The key here is the setup. You'll want to be able to set up the rubber band so it's ready to jump in less than half a second. When you get good at this, the move will be second nature and you'll even be able to talk to spectators as you perform the setup, which provides some misdirection. You can simply ask spectators if they've ever seen a jumping rubber band, and by the time they look at your hands again, you'll be ready to perform the trick.

You can also reverse the trick. With the rubber band on your first and second fingers, you can pull the band towards your pinky, insert all of your fingers, rest the band on your fingertips, and you're ready to open your hand to make the band jump back to the starting fingers.

The Jumping Rubber Band—Version 2

The effect becomes more baffling with version 2.

Effect

In this version, two rubber bands are placed on your fingers and they switch places. The trick uses the exact same setup as that used in the basic version, but you're setting up two rubber bands. And spectators who know the basic jumping rubber band will be baffled. There's seemingly no way that the rubber bands can quickly change places.

Materials and Requirements

Two rubber bands of different colors. The higher the contrast between the two colors, the better.

Skills

The ability to perform the basic "Jumping Rubber Band."

Preparation

This one can be done at any time and there is no preparation.

Performing the Trick

Wrap the rubber band around the ring and pinky fingers of your left or right hand. Wrap the second rubber band around your first and second fingers, as shown in Figure 7.5.

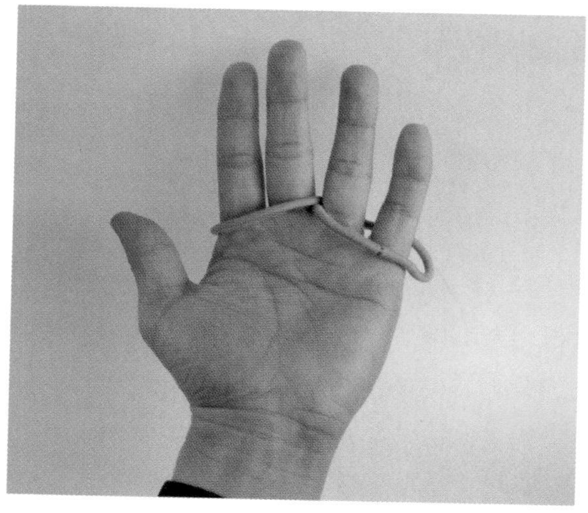

Figure 7.5
Wrap the first rubber band around your ring and pinky fingers and the second rubber band around your first and second fingers.

With the back of your hand facing spectators, use the first finger of your other hand to pull and stretch the first rubber band towards your thumb, as shown in Figure 7.6. This is the same move as that used in the basic "Jumping Rubber Band."

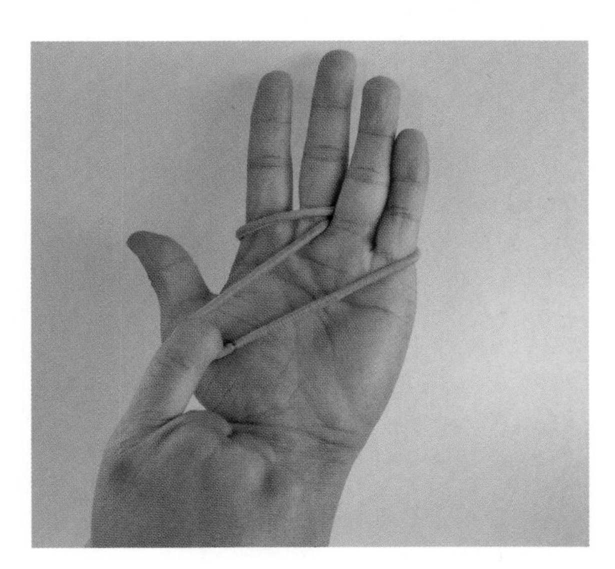

Figure 7.6
Pull the first rubber band towards your thumb.

While holding the first rubber band, use the second finger of your hand to pull the second rubber band towards your pinky finger, as shown in Figure 7.7.

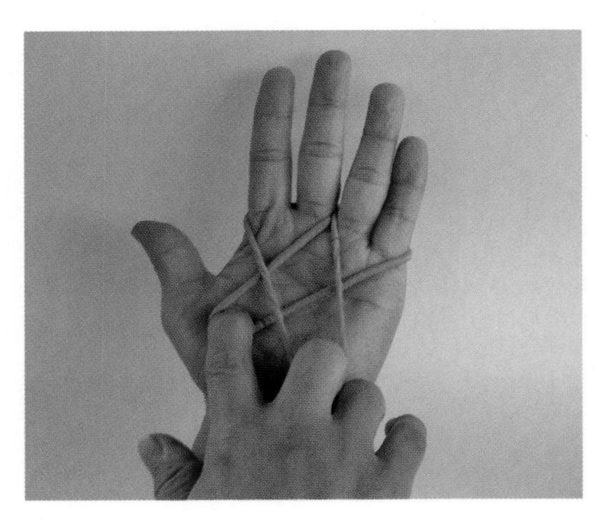

Figure 7.7
Pull the second rubber band towards your pinky finger.

You'll notice a diamond shape that's formed by the rubber bands in the center of your hand. Stretch this diamond with your other hand to widen it and insert all of your fingers into it, as shown in Figure 7.8.

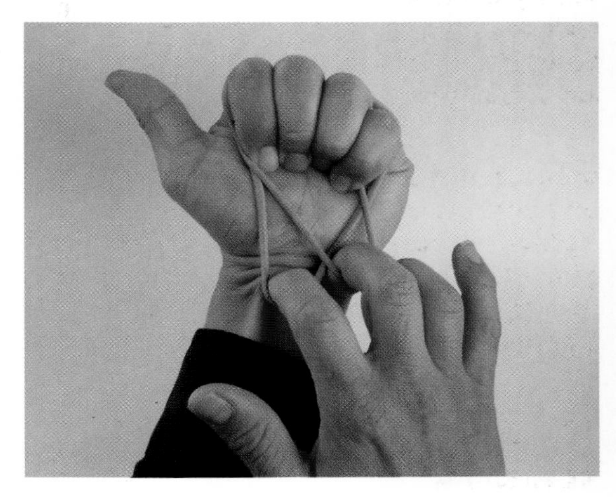

Figure 7.8
Insert your fingers into the diamond formed by the rubber bands in the center of your hand.

Rest the rubber bands on the tips of your fingers, as shown in Figure 7.9.

Figure 7.9
Rest the rubber bands on the tips of your fingers.

Open your hand to cause the rubber bands to jump through each other to the opposite set of fingers, as shown in Figure 7.10.

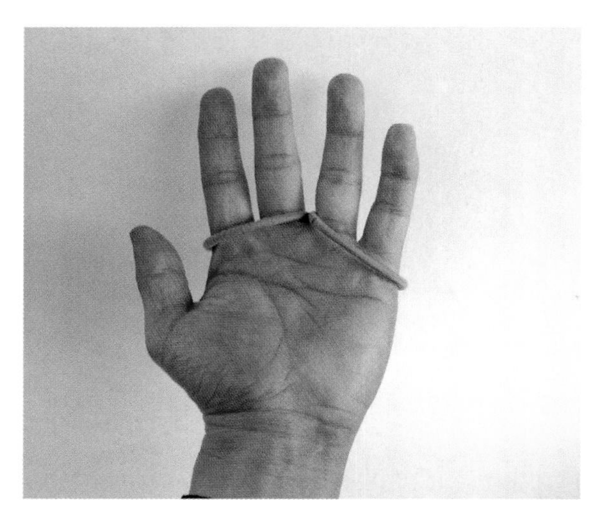

Figure 7.10
Open your hand to cause the rubber bands to jump through each other and to the opposite set of fingers.

The Jumping Rubber Band—Version 3

One more variation, and this is a baffling one.

Effect

One rubber band is wrapped around the tips of your fingers to create a barrier that prevents a second rubber band from jumping between your fingers. Despite the barrier, the rubber band still manages to jump in-between.

Materials and Requirements

Two rubber bands of different colors. You can use the same rubber bands that you used in the prior version.

Skills

The ability to perform the basic "Jumping Rubber Band."

Preparation

This one can be done at any time and there is no preparation.

Performing the Trick

Wrap the rubber band around the ring and pinky fingers of your left or right hand to set up the basic jumping rubber band, as shown in Figure 7.11.

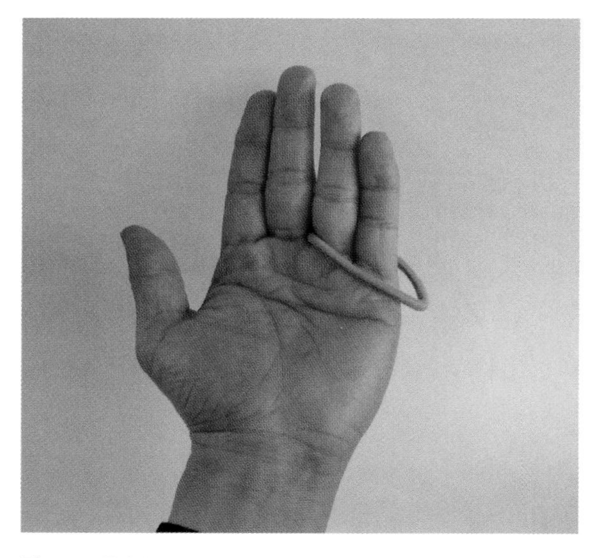

Figure 7.11
Wrap the rubber band around the ring and pinky fingers.

Using the second rubber band, wrap it across the tips of your fingers. You'll be wrapping the band around your pinky, giving the band a twist, wrapping the band around your third finger, giving the band a twist, and so on, until you've wrapped each finger and enclosed each with a twist, as shown in Figure 7.12.

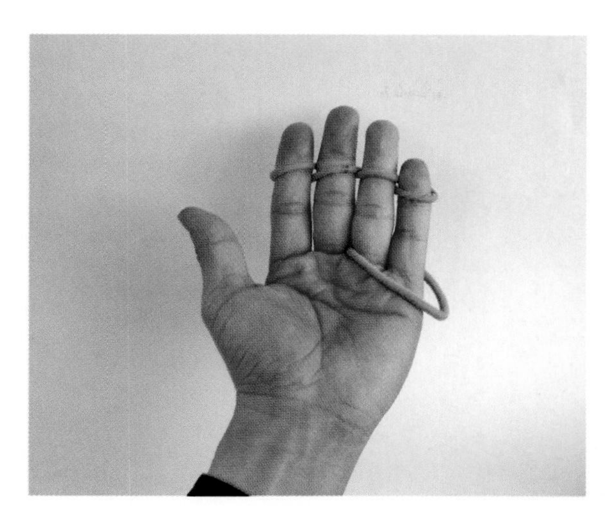

Figure 7.12
Wrap the second band around the tips of your fingers so you have a twist in-between each finger.

Finish by setting up the basic "Jumping Rubber Band" with the first band by pulling the band towards your thumb, inserting all of your fingers and resting the band on your fingertips, as shown in Figure 7.13.

Figure 7.13
Set up the basic "Jumping Rubber Band," as explained earlier.

By opening your hand, you make the first band jump between your fingers, despite the blocking rubber band, as shown in Figure 7.14.

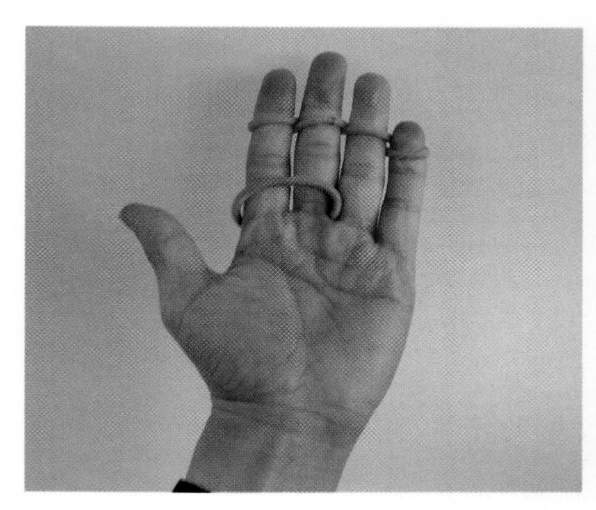

Figure 7.14
The rubber band still jumps to the other fingers despite being blocked by the second band.

While this looks impossible, the band actually flips over the blocking band. While this is hard to explain, you can see how this works by performing the trick slowly. Set up the trick and then hold the band with your other hand as you slowly open your hand. You'll see how the jumping band gets by the blocking band.

Rubber Band Through the Thumb

A rubber band seemingly passes through your thumb.

Effect

You wrap a rubber band twice around your right thumb and are able to make the band pass through your thumb to free it. This one is harder than the jumping rubber band sequences. Expect to experiment and practice a bit to get this one right.

Materials and Requirements

One rubber band. This can be the same band that you've used in the other rubber band tricks.

Preparation

This one can be done at any time and there is no preparation.

Performing the Trick

Hold the rubber band around the left first and second fingers and insert the right thumb from the opposite direction, as shown in Figure 7.15. The left first and second fingers enter the band from the top and the right thumb enters from the bottom.

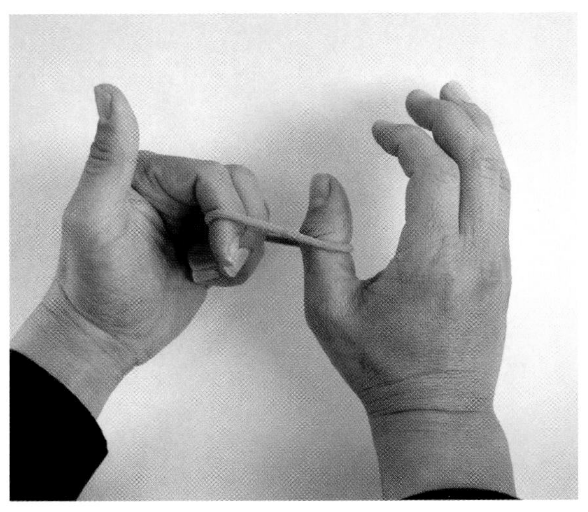

Figure 7.15
Hold the band with the left fingers on one side and the right thumb in the other. Notice how the fingers and thumb are entering from opposite sides into the band.

Using the first finger of the right hand, clip the closest strand of the rubber band, as shown in Figure 7.16.

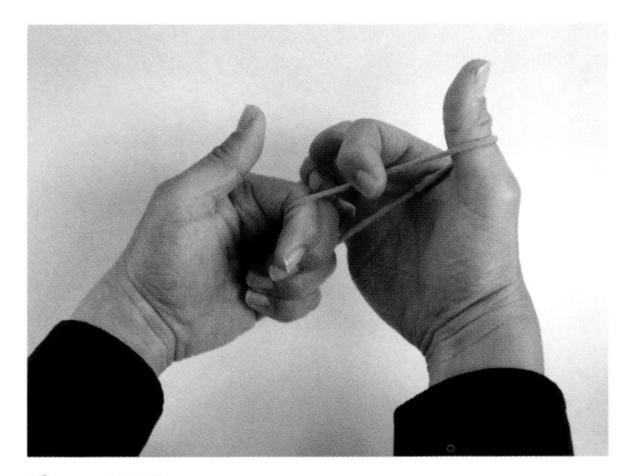

Figure 7.16
Using the right first finger, clip the closest strand.

Pull the clipped strand back towards the right hand and turn the fingers of the left hand away from you. The left hand is turned palm down. You'll end up with an hourglass shape in the rubber band, as in Figure 7.17.

Figure 7.17

Pull the strand back with the first finger and pivot the left hand fingers away. You should see an "hourglass" in the center.

Insert the right thumb into the leftmost half of the hourglass through the top, as shown in Figure 7.18.

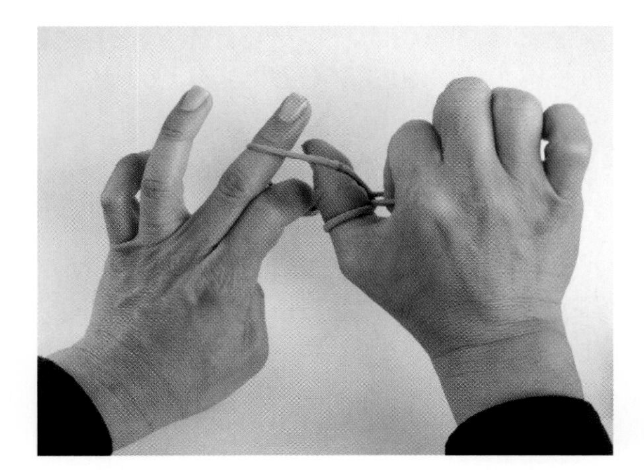

Figure 7.18

Insert the right thumb into the left half of the "hourglass."

Wrap the "x," the center of the "hourglass" over the loop that's clipped by the right first finger. With the left hand, pull the band so it's taut and holds the band in place around your thumb. It now looks as if you've wrapped the rubber band around your right thumb twice.

When done right, you'll be able to let go of the right first finger and the thumb will stay wrapped. If the rubber band slips off of your thumb, try holding more tension in the band by pulling with the left fingers. Also, make sure that the "x" of the hourglass lays over the loop that's originally clipped by the right first finger behind the thumb. This secures the band behind the thumb.

To make the penetration more convincing, touch your right second finger against the tip of your right thumb. It's also more dramatic if you insert your left thumb into the band to position the band vertically. You'll be holding the band as shown in Figure 7.19.

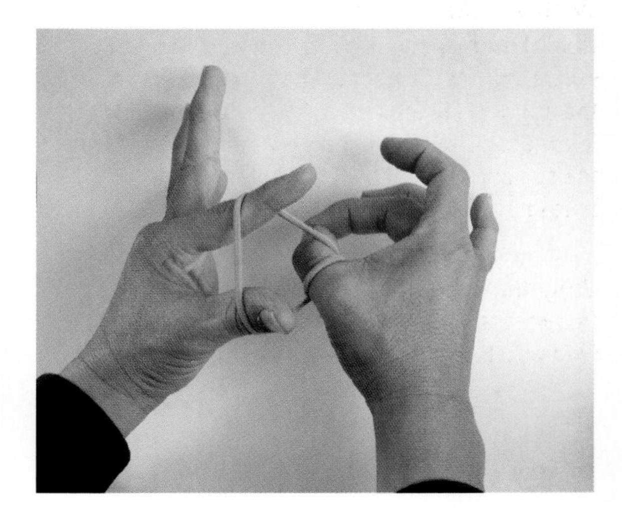

Figure 7.19

The final hand and rubber band position before the band passes through your thumb.

By shifting your right thumb, you cause the band to slip off of your thumb. The rubber band appears to pass through your thumb, as shown in Figure 7.20.

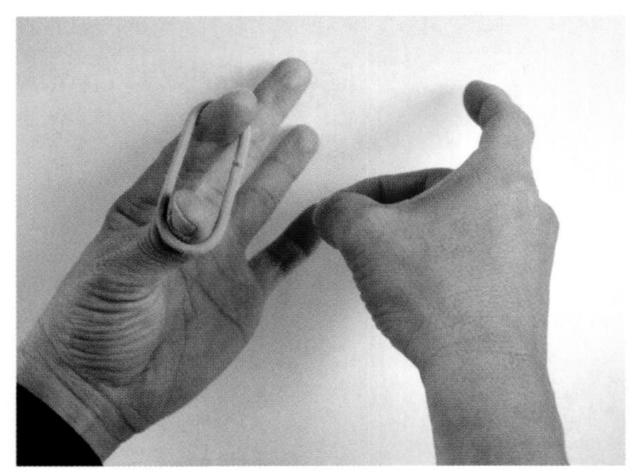

Figure 7.20
The rubber band appears to pass through the thumb.

If you can set this up but the band keeps slipping off of your thumb, you've got the right setup. You just haven't gotten the right tension. You'll need to experiment until you get the setup and tension correct.

Note that when performing this effect, you hold your hands down, somewhat below the gaze of your audience. This way, they don't see the loop around your thumb.

The Broken and Restored Rubber Band

Effect

You break a rubber band—spectators will actually hear the band break. But after placing the broken band into your hand, you completely restore it.

Materials and Requirements

One rubber band. This can be the same band that you've used in the other rubber band tricks. However, a thinner rubberband is generally preferable as it makes the illusion more convincing.

Preparation

This one can be done at any time and there is no preparation.

Performing the Trick

Hold the rubber band around the first, second, and third fingers of the right hand, as shown in Figure 7.21.

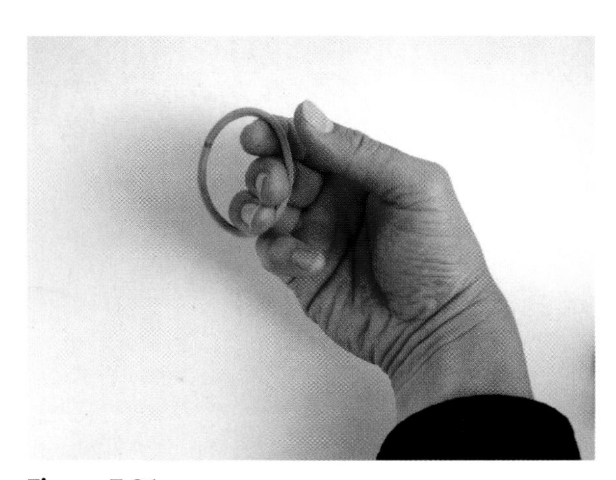

Figure 7.21
Hold the band around the first three fingers of your right hand.

Reach with the first finger of the left hand to clip and pull half of the band, as shown in Figure 7.22.

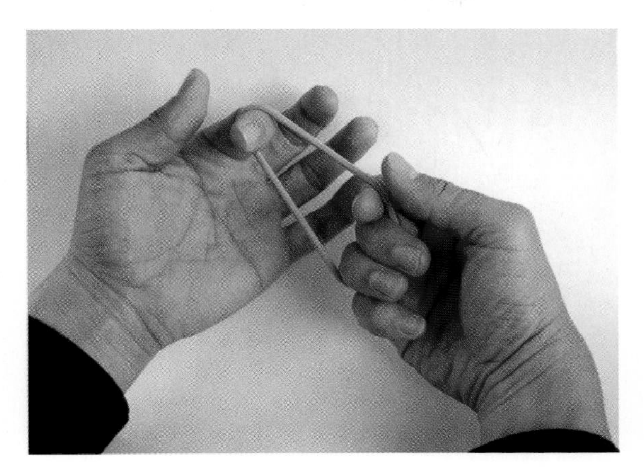

Figure 7.22
Pull the band with the first finger of the left hand.

Using the second and third fingers of the left hand, push the top strand down. The rubber band is double-backed and in the general shape of the letter "V," as shown in Figure 7.23.

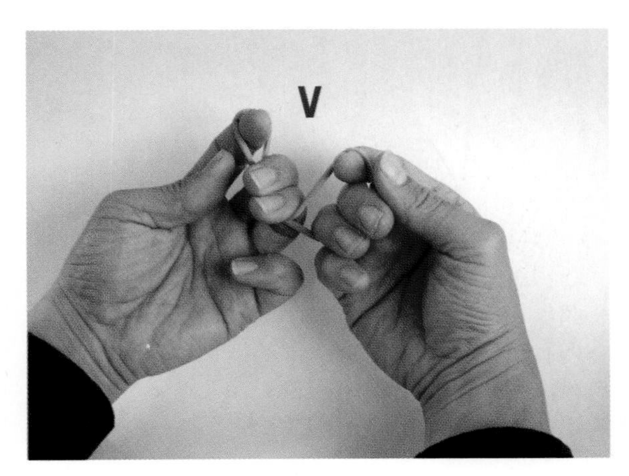

Figure 7.23
Push down with the second and third fingers of the left hand. The rubber band is doubled-back and in the general shape of the letter "V."

Holding the band down with the third finger of the left hand, the thumb and middle finger grab and pinch the band just below the first left finger, as shown in Figure 7.24.

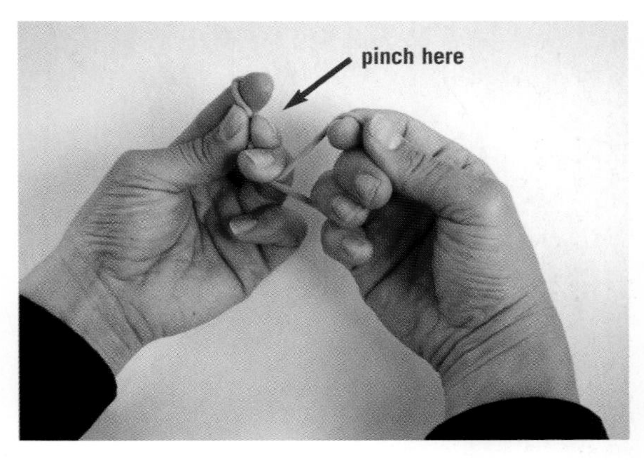

Figure 7.24
Pinch the band with the left thumb and third finger.

The left first finger slips out as the left thumb and second finger hold the band. There will be a tiny loop above the thumb and third finger that was left by the first finger, as shown in Figure 7.25.

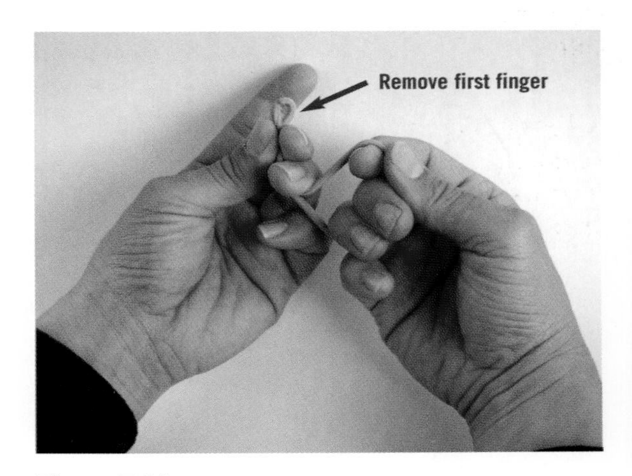

Figure 7.25
The first finger slips out while the left thumb and second finger maintain their hold on the band.

The left thumb and second finger bring the loop to the right, through the right portion of the band, where it's held by the right thumb and first finger, as shown in Figure 7.26. Note that the right thumb has been brought from its location outside the loop in Figure 7.25 to inside the loop, as shown in Figure 7.26.

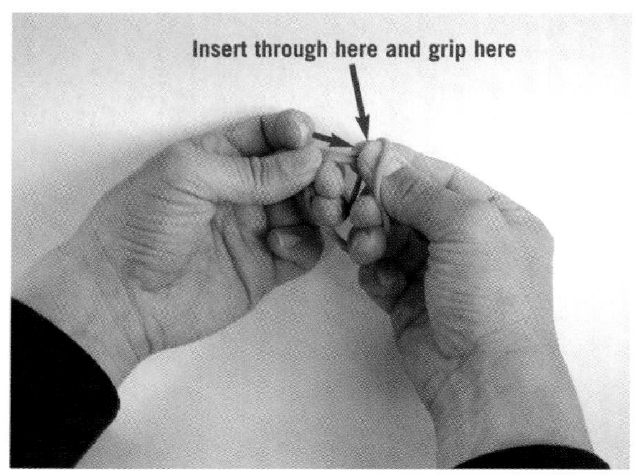

Figure 7.26
The leftmost portion of the band is delivered through the rightmost portion of the band to the right thumb and first finger.

The right thumb and first finger maintain their grip on the newly received loop, but allow the loop on top to slip off, as shown in Figure 7.27.

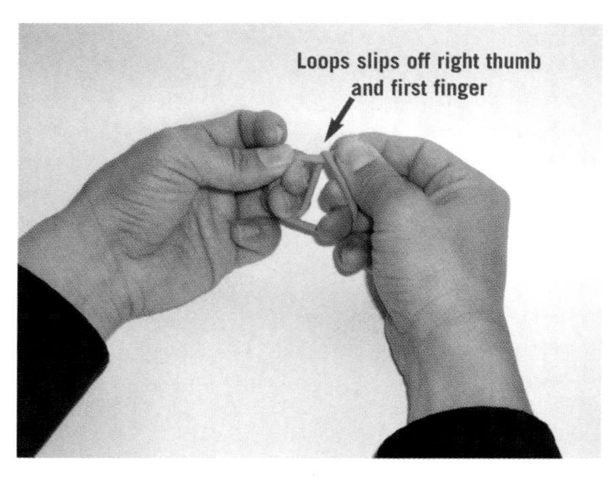

Figure 7.27
The right thumb and first finger allow the loop on top of it to slip off.

The right second finger reaches and hooks the single strand in the middle of the "U" and pulls it to the right, as shown in Figure 7.28.

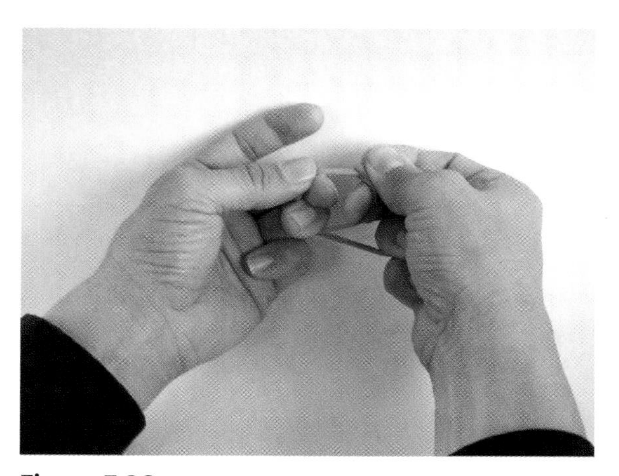

Figure 7.28
The right second finger reaches for the center strand and pulls it to the right.

The right third finger reaches and hooks the center strand and pulls it to the right. You're creating a loop that consists of a double-strand of rubber band that's held together by the right thumb and first finger, as shown in Figure 7.29.

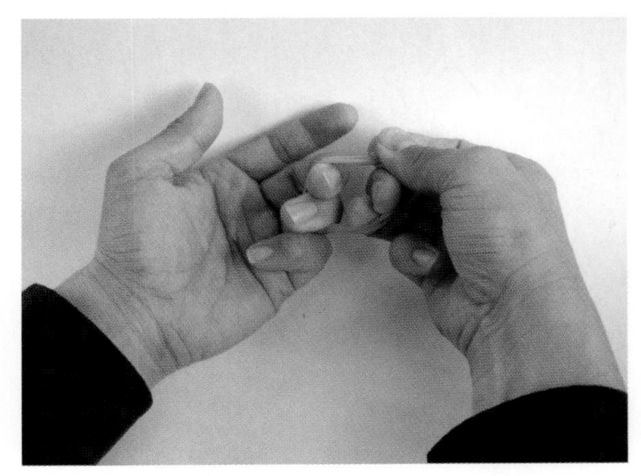

Figure 7.29
The right third finger reaches for the center strand and pulls it to the right.

Fully stretched, the double-thickness band looks like a normal band. Stretching the band helps to hide the fact that the perimeter consists of two thicknesses of bands, as shown in Figure 7.30. Again, we are demonstrating here with hair bands for visibility.

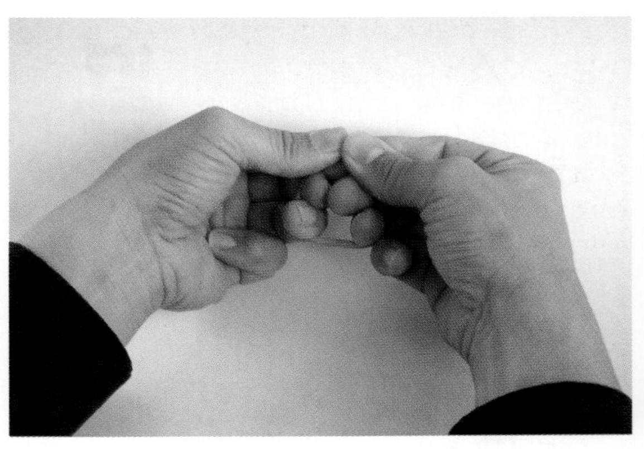

Figure 7.30
The final configuration before the "break."

Using the left thumb and first finger, grip the top double-thick strand of the band near the right thumb and first finger.

Reality Check

You now appear to be holding a rubber band, but it's actually a band that's been doubled over and closed again at the point where the left thumb and first finger and right thumb and first finger meet. There's a "break" in the band at this point.

Holding the respective ends, separate and appear to break the band. By forcefully rubbing the ends of the rubber band against each other, you can make a noise that sounds like a snapping rubber band. Be sure to stretch the length of band between the hands to hide the fact that this segment consists of two bands. This is shown in Figure 7.31.

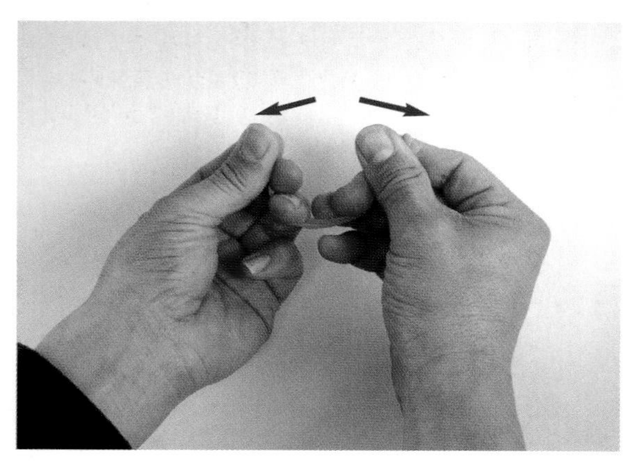

Figure 7.31
Appear to break the band by forcefully separating the two "ends" of the loop.

Place the entire band into your left palm and release the right and left fingers and thumbs under cover of the left fingers. Open your hand to show that the rubber band is now whole again and your hands are completely empty, as shown in Figure 7.32.

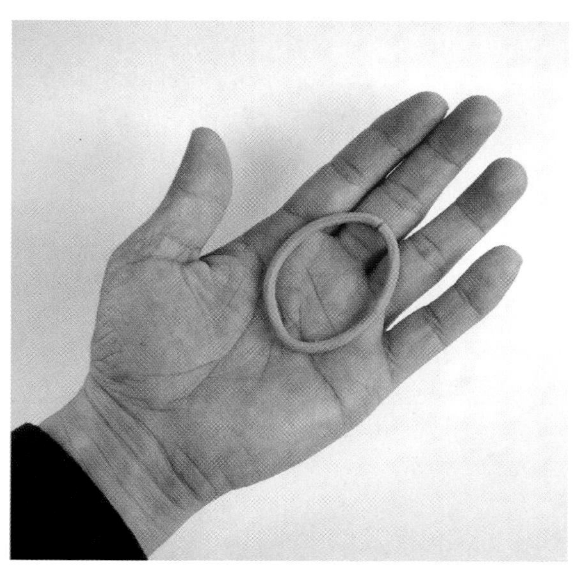

Figure 7.32
The band is restored.

Coin Magic

EVERYONE LOVES MONEY AND you can almost always scare up a few coins to perform a quick trick or two. Here, we offer a basic course in coin sleight-of-hand that includes some means to vanish and reproduce coins and we teach short routines that apply what you've learned.

What Types of Coins Should You Use?

Magicians tend to favor half-dollars and silver dollars. Personally, I use large, old-style silver dollars in my close-up work. The key here is visibility. The larger coins are easier for spectators to see and follow. For the coin tricks in this chapter, you can use quarters or any size coin that's comfortable for you to work with.

Basic Coin Sleights

One of the fundamental techniques in magic is being able to hold an object in your hand without spectators realizing that there's something there. Magicians have a term for this. We call this "holding an object in your hand without spectators realizing that there's something there." (Just kidding.) Of course, the technique is known as "palming."

There are techniques in magic for palming all kinds of objects, and we're going to begin your palming education by showing you how to palm a coin and then explain some fundamental moves.

The Finger Palm

Using the finger palm, you can hold coins of various sizes and your hand will still look natural and not draw attention to itself. With the finger palm, you secretly hold a coin between the base of your second and third fingers and the first knuckle of your finger, as shown in Figure 7.33.

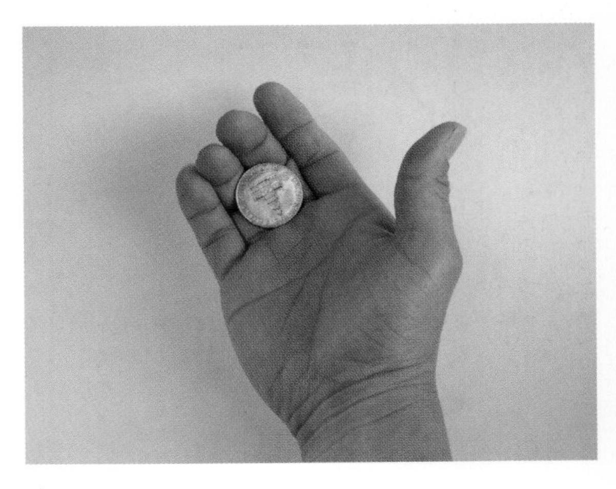

Figure 7.33
The position of a coin that's held in a finger palm.

By adjusting the curl in your fingers, you can accommodate a wide variety of coins and securely hold them.

The goal is to make the hand look natural. If you stiffen your hand or hold the coin too tightly, it will draw attention to itself and appear suspicious. Even though it's holding a coin, you want your hand to look completely relaxed, as shown in Figure 7.34.

Figure 7.34
The relaxed position of the hand while finger palming a coin.

The finger palm is among the easiest palms to learn and master. Try to practice the move by placing the coin into the finger palm position and then adjusting the tension in your hand until it looks natural. Once you've got a natural appearance, try to achieve this look each and every time you palm a coin.

Another aspect to practice is to get the coin into finger palm without using the other hand. Many times, as you'll soon see, the coin pretty much lands in your hand and you'll have to palm the coin without the assistance of the other hand.

The key here is to try and make the coin land in the position to be readily finger palmed, or by using your thumb, quickly get the coin into position before the palm. One more thing: it's a good idea to practice finger palming with both hands.

It's a good idea to experiment with the coins that work best for you. Generally, newer coins with sharper edges are easier to palm. Also, your ability to easily palm coins will depend on the condition of your skin. Too moist and coins can become slippery. And too dry, the coins can be difficult to hold. To adjust for variations in your skin, you can try using skin lotion.

Hint

To master palming, it's a good idea to keep a coin in your pocket and palm it while you're attending to your day to day routines. Soon, you won't even have to think about that coin in your hand, which is what you want. If you're not thinking about holding the coin, you won't draw attention to your hand.

The Shuttle Vanish

A fundamental move in magic is pretending to place an object in one hand, but secretly keeping it or palming it in the other. When you open the receiving hand, it appears that the object has disappeared or "vanished," but it was actually retained in the other hand.

The term "vanish" refers to a host of moves designed to make it look as if you've placed a coin elsewhere. In a shuttle vanish, you'll show a coin in one hand that is ready to be finger palmed—as explained in the previous section, the coin is resting against the base of the second and third fingers and the fingers are ready to be curled in to secretly grip the coin, as shown in Figure 7.35.

Figure 7.35
The coin is positioned in the hand to be finger palmed.

You'll be turning your hand over as if you're going to drop the coin into the other hand, but you will actually retain it in finger palm position, as shown in Figure 7.36.

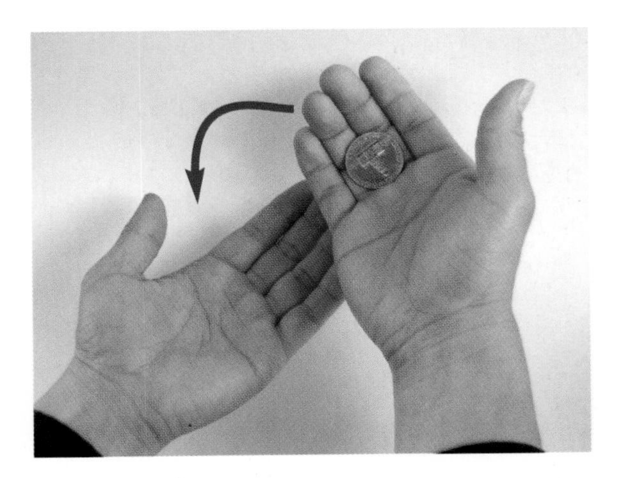

Figure 7.36
As you turn your hand, you'll finger palm the coin.

At the same time that you're finger palming the coin, your other hand closes as if it actually received the coin. Put together, it will look as if you are dropping a coin into your other hand, but you're retaining it in finger palm, as shown in Figure 7.37.

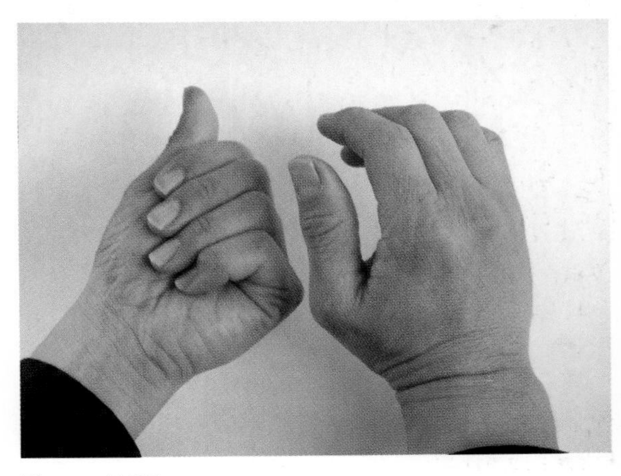

Figure 7.37
The receiving hand closes as if it received the coin.

A good way to develop your shuttle vanish is to work in a mirror and actually drop a coin from one hand to the other. Carefully observe how this looks when you actually transfer the coin.

Now, practice executing the shuttle vanish in a mirror and try to make it look as if you've actually transferred the coin. When you can perform a shuttle vanish so you can't tell by watching, you've mastered the move. One more thing, try to master the move with both hands so you can vanish a coin in either direction.

The French Drop

A second vanish that we'll teach you is known as the French drop. Instead of tossing a coin from one hand to the other, as in the shuttle vanish, in the French drop, you appear to take a coin from one hand when you're actually leaving it behind.

Hold the coin at the tips of the second and third fingers and the thumb, as shown in Figure 7.38.

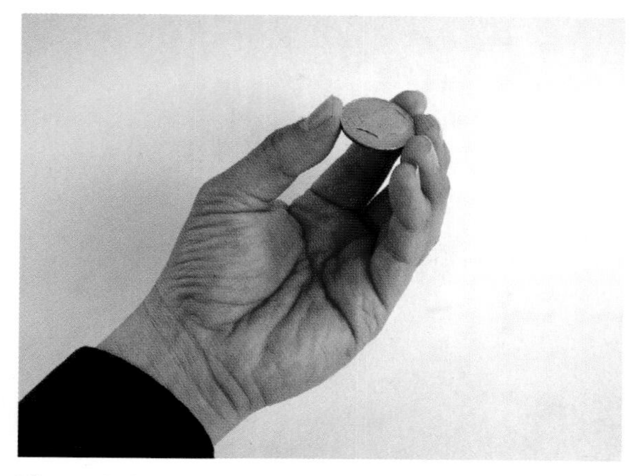

Figure 7.38
Preparing for the French drop.

With your other hand, "grab" the coin with the thumb underneath and the fingers on top, as shown in Figure 7.39.

As your grabbing hand appears to enclose the coin, you release your thumb and allow the coin to fall into your hand, as shown in Figure 7.40.

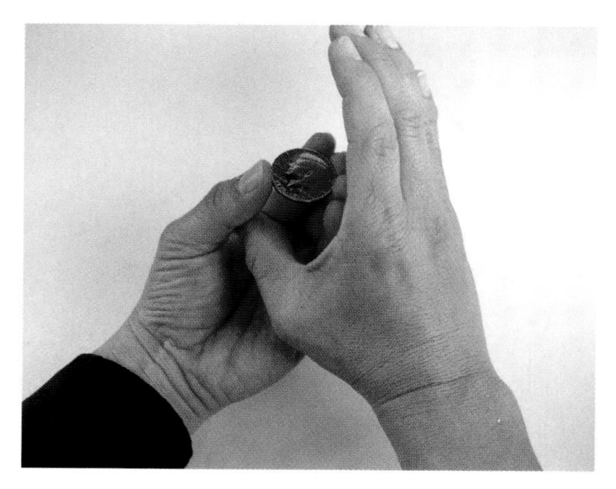

Figure 7.39
The second hand grabs at the coin.

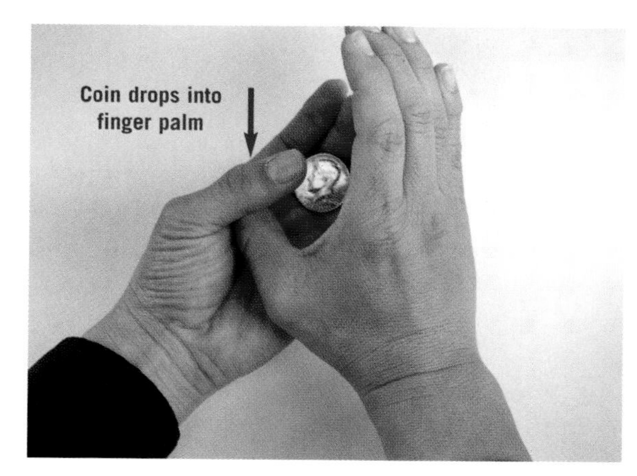

Figure 7.40
The coin is released and falls into finger palm. Note that this is an exposed view.

Finger palm the coin. The grabbing hand closes and appears to take the coin away. As the grabbing hand moves away, the other hand with the finger palmed coin turns down, as shown in Figure 7.41.

Figure 7.41
As the grabbing hand moves away, the coin is finger palmed in the first hand and the hand is turned down.

The tendency for most beginning magicians is to quickly move the hand with the palmed coin away. However, this draws attention to the hand with the palmed coin. It's best to simply turn over your hand and freeze it in place and draw your own eyes and attention to the hand that is supposed to be holding the coin.

When performing vanishes such as the French drop, you have to act as if your other hand actually took the coin to draw attention away from the other hand. And when your attention and that of your audience is drawn to your empty hand, you can casually lower your other hand with its palmed coin and move it away.

As with the shuttle vanish, try to master the French drop so you can perform it with both hands.

Producing a Coin

Once you've vanished a coin, you have to make it reappear. To do this, you'll bring the coin out of finger palm position to the tips of your finger.

With a coin finger palmed, take your thumb and place it under the bottom edge of the coin, as shown in Figure 7.42.

Figure 7.42
Your thumb positions itself below the finger palmed coin.

Using your thumb, push the coin out of finger palm position and to the tips of the finger, as shown in Figure 7.43.

Figure 7.43
Push the coin out of finger palm and to the tips of the fingers.

You want to make it look as if you've plucked the coin from the air, seemingly out of nowhere. To do this, reach your hand into the air as if the coin is sitting, suspended invisibly in a specific location, and as your hand reaches forward, quickly produce the coin at your fingertips as you pull your hand back. The coin should look as if it materialized at your fingertips.

Another method is to produce the coin from behind an object such as your elbow or knee, or from an empty cup or beneath a place mat. To do this, reach your hand behind your elbow and while it's momentarily hidden, bring the coin out to your fingertips and then show spectators the coin.

The Ten-Count Coin Trick

With your newfound coin techniques, here's a trick where you count to ten and make the magic happen. Because you're counting, you create a rhythm that carries spectators along, and a bit of misdirection also helps.

Effect

You cause two coins to magically join together in your hands as you count to ten. Magicians refer to tricks of this type as "coins across" effects.

Materials and Requirements

Two coins (these can be the same or different).

You'll need to perform this one on a table (a table that is covered with a table cloth makes it easier to pick up the coins, but you can perform this on any flat surface, including the ground).

Skills

Finger palm.

Shuttle vanish.

Preparation

None.

Performing the Trick

Lay out the coins as shown in Figure 7.44.

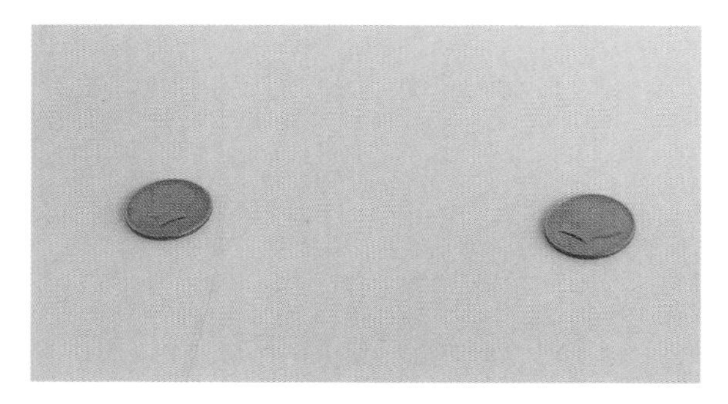

Figure 7.44
The starting position of the coins.

Call out "one" and cover the leftmost coin with your left hand, as shown in Figure 7.45.

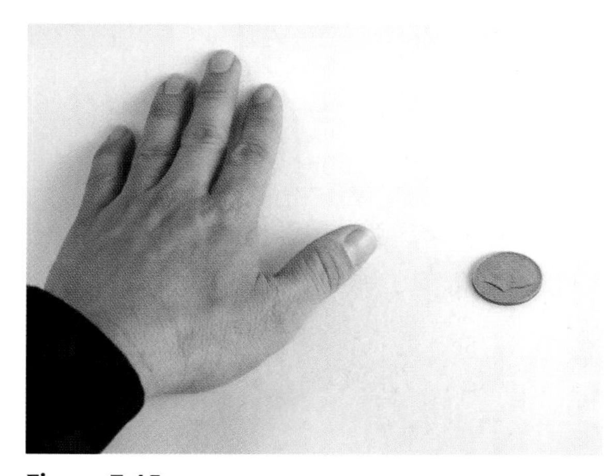

Figure 7.45
"One"—cover the leftmost coin with your left hand.

Call out "two" and cover the rightmost coin with your right hand, as shown in Figure 7.46.

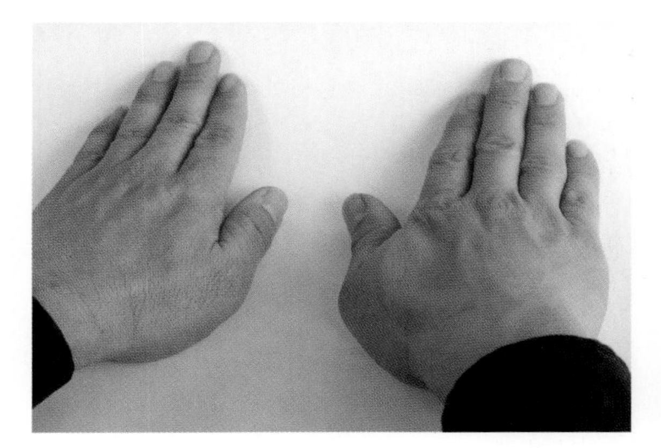

Figure 7.46
"Two"—cover the second coin with your right hand.

Call out "three" and turn your left hand over, as shown in Figure 7.47.

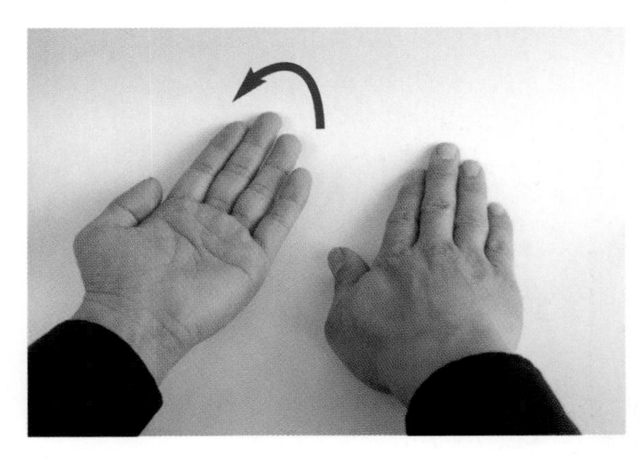

Figure 7.47
"Three"—turn your left hand over.

Call out "four" as you pick up the rightmost coin and hold it ready to be finger palmed, as shown in Figure 7.48.

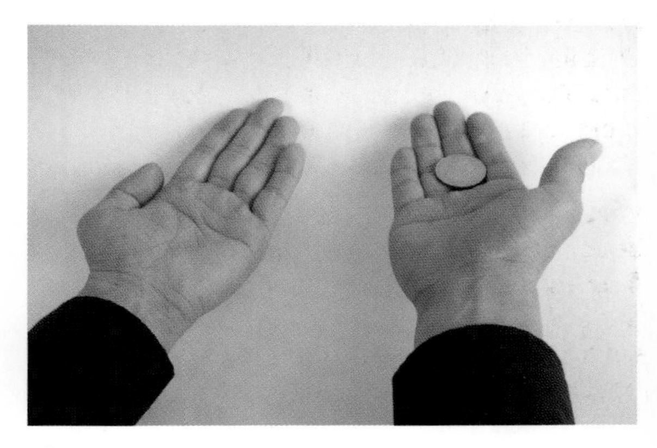

Figure 7.48
"Four"—pick up the second coin and prepare it for finger palm.

Call out "five" and perform a shuttle vanish with the right hand, as shown in Figure 7.49.

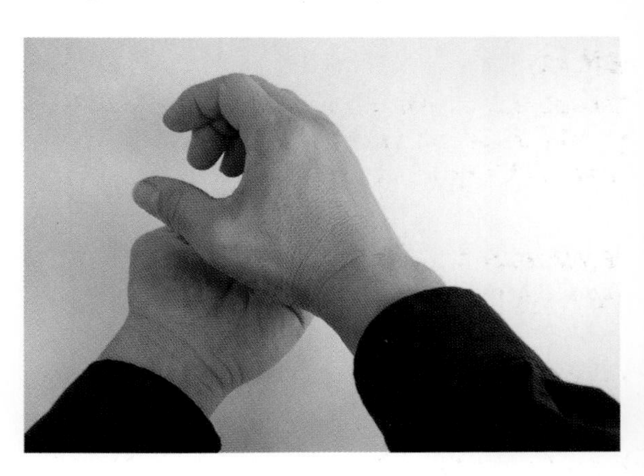

Figure 7.49
"Five"—execute a shuttle vanish.

Reality Check

It looks as if you've placed the coin in your left hand, but it's actually finger palmed in your right hand.

Call out "six" and lift your left hand to reveal the coin under it. Note the strong, built-in misdirection here. Before the spectator has an opportunity to think about whether you've transferred the coin or not, you reveal the second coin that makes spectators concentrate on something new. This is shown in Figure 7.50.

Figure 7.50
"Six"—lift your left hand to reveal the first coin.

Call out "seven" and pick up the leftmost coin with your right hand. You'll have to practice holding a coin finger palmed in the right hand and picking up a second coin.

Hold the second coin at the fingertips with the first or second finger and the thumb. It's important that you learn to grab and hold the second coin so it doesn't hit the first coin and make a sound. This is shown in Figure 7.51.

Figure 7.51
Grab the first coin with your right hand.

Call out "eight" and allow the two coins to hit each other in the right hand so that they make a sound, as shown in Figure 7.52.

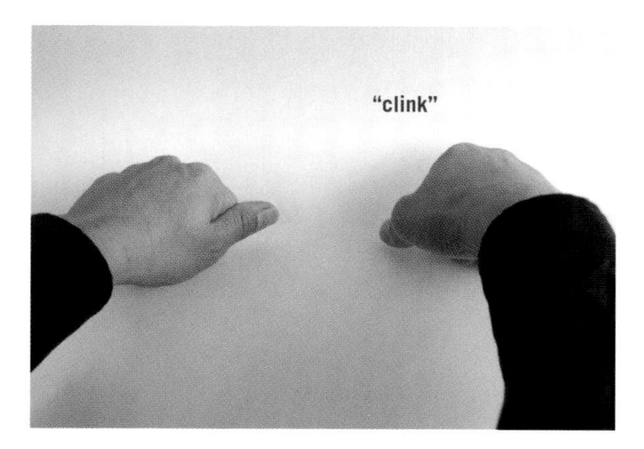

Figure 7.52
"Eight"—allow the coins in your right hand to hit and make a sound.

Call out "nine" and open your left hand to show that it's empty, as shown in Figure 7.53.

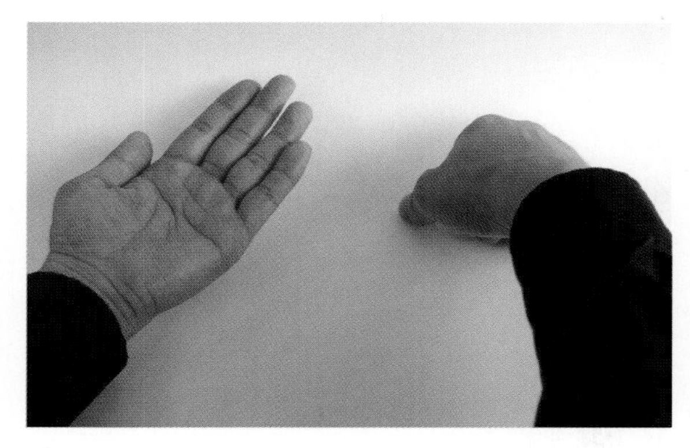

Figure 7.53
"Nine"—open your left hand.

Call out "ten" and show your right hand with two coins. You've made a coin disappear from one hand and join the other one, as shown in Figure 7.54.

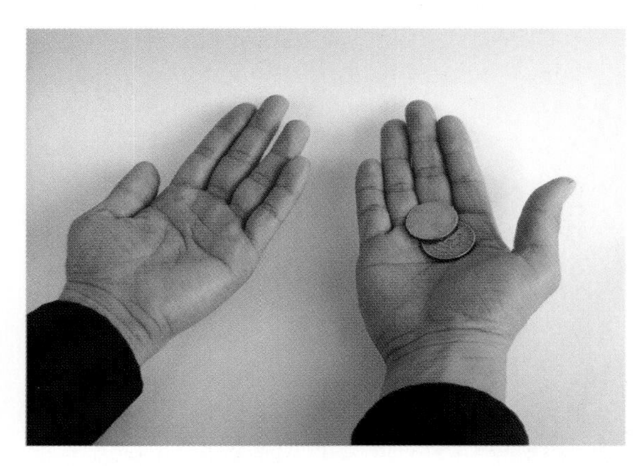

Figure 7.54
"Ten"--Open your right hand.

To make this trick work, aim to keep a constant cadence when you count through the numbers and try not to alter the speed in certain places. By keeping a constant cadence, you'll walk spectators through the trick, and the numbers help spectators stay with the flow.

Don't repeat this trick no matter how many times spectators ask (and they will).

Coin Through Handkerchief

Effect

You cause a coin to magically pass through the fabric of a handkerchief.

Materials and Requirements

A coin.

A handkerchief, bandanna, or cloth napkin.

Secret

You surreptitiously bring the coin to the outside of the handkerchief before spectators realize anything has happened.

Preparation

None.

Performing the Trick

Hold the coin in the left hand and using the right hand, drape the handkerchief over the left hand, as shown in Figure 7.55.

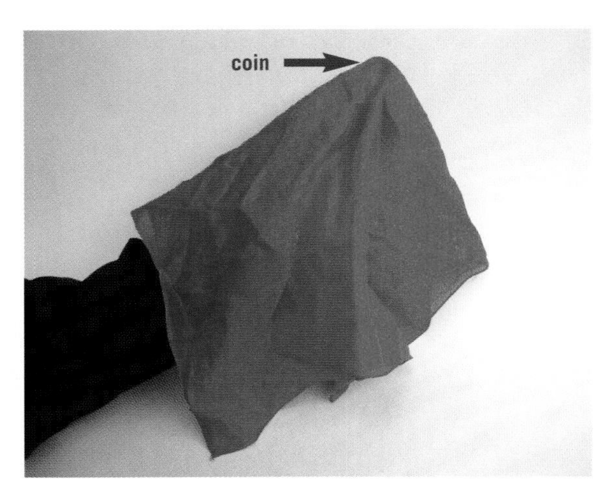

Figure 7.55
Drape the handkerchief over the coin.

Using your right hand, pinch the coin through the handkerchief and fold the coin down (through the fold) toward you. You've folded the coin through the fabric and there are now two layers of cloth between your fingers and the coin, as shown in Figure 7.56.

Figure 7.56
Pinch the coin in the cloth and fold it toward yourself.

Lift up the front edge of the handkerchief (the edge farthest from you) to show spectators that the coin is still under the handkerchief, as shown in Figure 7.57.

Figure 7.57
Lift the front edge of the handkerchief to show that the coin is under the handkerchief.

Grab not only the front edge of the handkerchief but the back edge that's nearest you. Pull both edges and allow them to fall forward, as shown in Figure 7.58. You will now be holding the coin from above.

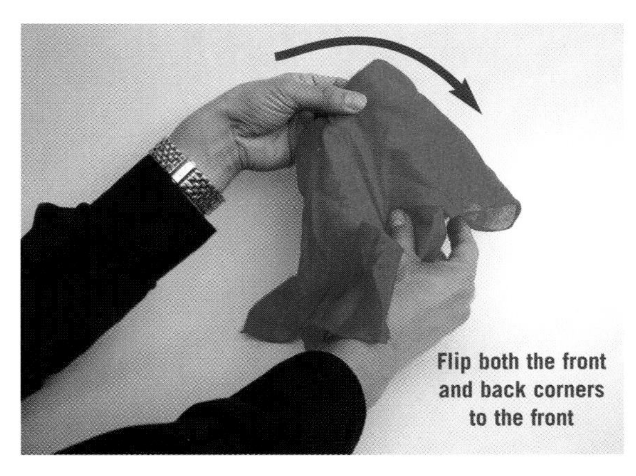

Figure 7.58
Grasp both the front and back edges and pull them forward.

Reality Check

By pulling both edges of the handkerchief forward, you have secretly brought the coin to the top side of the handkerchief, even though it appears to still be underneath. The coin is actually within a fold at the top center of the handkerchief.

Using your other hand, twist the handkerchief, as shown in Figure 7.59.

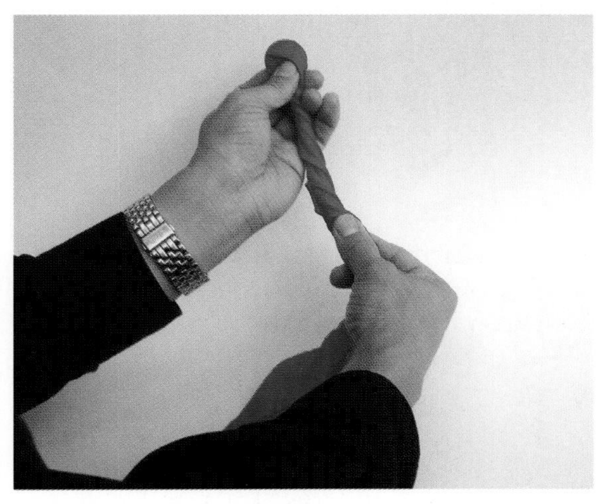

Figure 7.59
Twist the handkerchief.

Allow the coin to pop up through the top, as if it passed through the handkerchief, as shown in Figure 7.60.

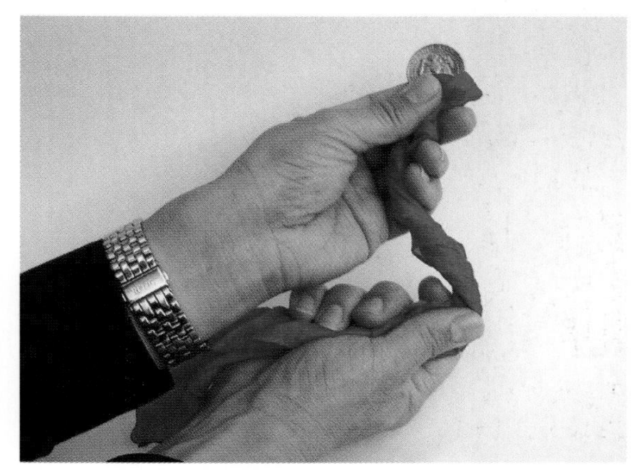

Figure 7.60
Allow the coin to pop up out of the top.

You can allow spectators to examine everything.

Coin Through Handkerchief and Finger Ring

Effect

You cause a coin to magically pass through the fabric of a handkerchief as well as through a finger ring that has a diameter too small to allow the passage of the coin.

Materials and Requirements

A coin.

A handkerchief or bandanna. (The thinner the material the better. You're going to have to twist the fabric and thread it through a finger ring.)

A finger ring, which can be borrowed. The coin should not be able to pass through the finger ring. The larger the diameter of the finger ring, the better.

Secret

You surreptitiously bring the coin to the outside of the handkerchief before spectators realize anything has happened.

Preparation

None.

Performing the Trick

Allow spectators to examine the coin, ring, and handkerchief.

Execute the "Coin Through Handkerchief" as explained in the previous section until you get to the point where the coin is secretly brought to the top of the handkerchief and you've twisted the handkerchief, as shown in Figure 7.61.

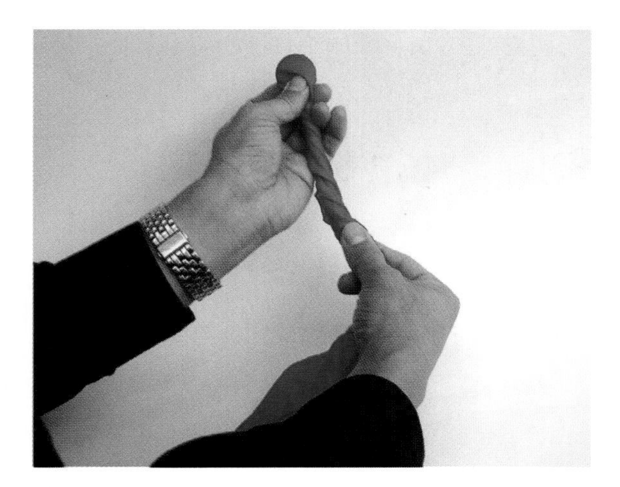

Figure 7.61
Perform "Coin Through Handkerchief" until the point where you're twisting the handkerchief.

While holding the coin, thread the twisted handkerchief through the finger ring until the ring rests up against the coin, as shown in Figure 7.62.

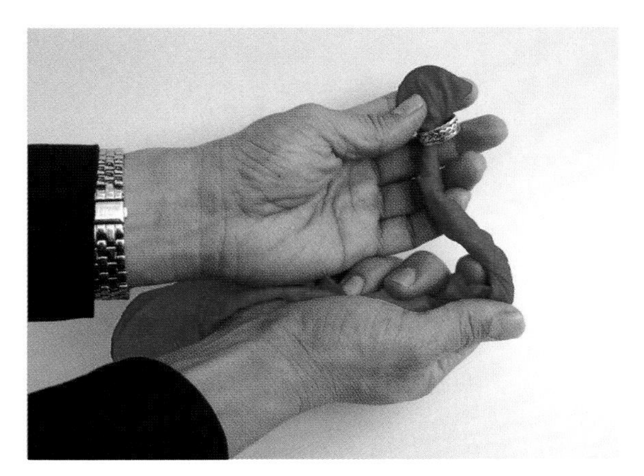

Figure 7.62
Slide the finger ring onto the twisted handkerchief.

While holding the coin through the fabric, find the four corners of the handkerchief and allow spectators to hold them, creating something of a makeshift table, as shown in Figure 7.63.

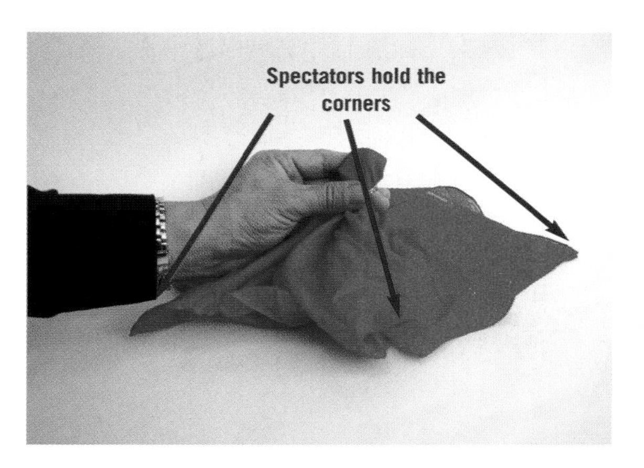

Figure 7.63
Have spectators hold the corners of the handkerchief.

Cover the coin with your hands and work the fabric open to secretly remove the ring. If you need to, ask the spectator for some slack so you have fabric to work with. Work out the ring and show it to spectators, as shown in Figure 7.64.

Figure 7.64
Cover the coin with your hands and work out the coin and ring. Show the ring first.

Even though the coin is already free, act as if you are still working on the coin and then bring out and show the coin, as shown in Figure 7.65.

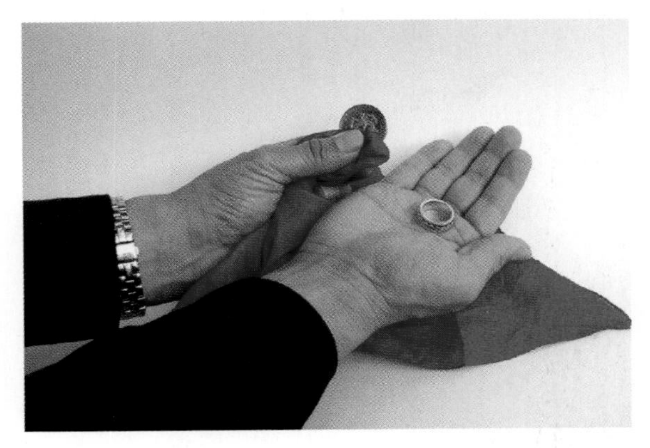

Figure 7.65
After some "work," bring out the coin.

Drop the coin and ring into the now taut handkerchief so spectators can examine them.

Coin Routine: Elbow to Elbow

Here's a short coin routine that you can perform using the sleight-of-hand moves that you learned earlier.

Effect

A coin repeatedly moves from elbow to elbow. At the end, you perform the ten step coin transposition.

Secret

It's all in the hands and moves.

Skills

French drop.

Finger palm.

Shuttle vanish.

Materials and Requirements

Two coins.

Preparation

None.

Performing the Trick

Bring out two coins and place them onto the table.

Pick up one of the coins with your right hand and prepare it for the French drop. Perform the French drop, as shown in Figure 7.66.

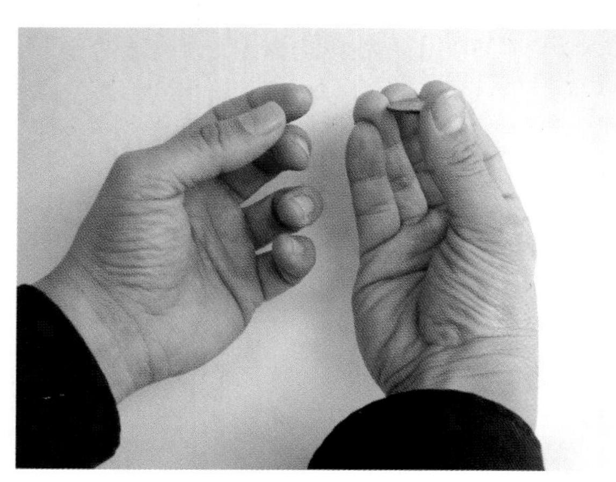

Figure 7.66
Pick up the coin with the right hand and prepare for the French drop.

Immediately cross your arms and move your closed left hand to your right elbow and with a throwing motion, pretend to place the coin into your elbow, as shown in Figure 7.67.

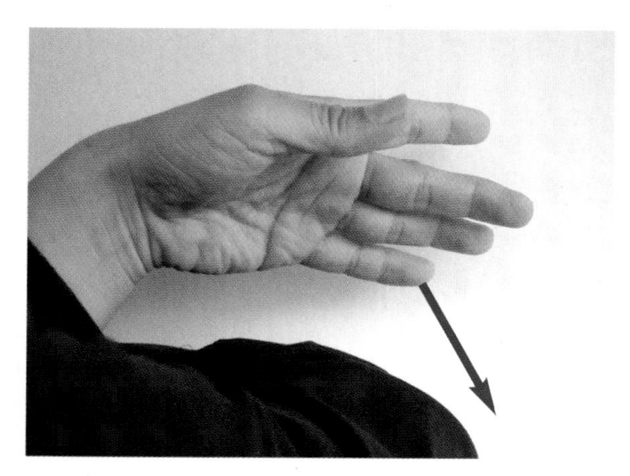

Figure 7.67
Throw the coin into your elbow where it disappears.

With your right hand, immediately produce the coin from your left elbow, as shown in Figure 7.68.

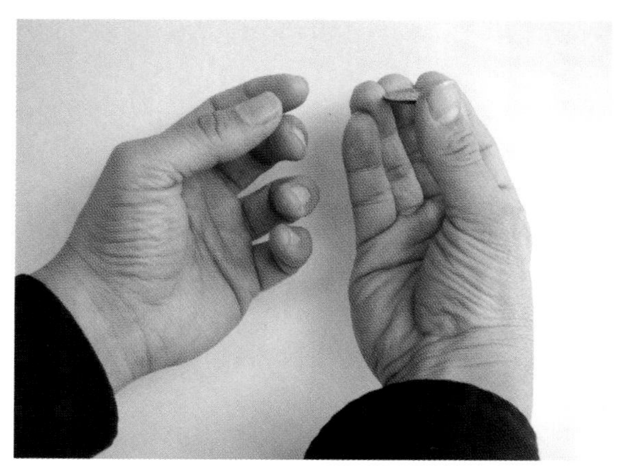

Figure 7.68
Produce the coin with your right hand from your left elbow.

Bring your right hand forward and show the coin openly in your hand. As you move your hand forward, get the coin into pre-finger palm position. Perform a shuttle vanish, as shown in Figure 7.69.

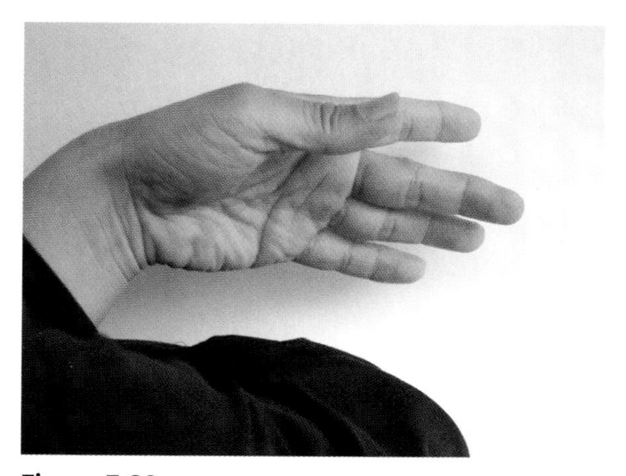

Figure 7.69
Perform a shuttle vanish.

Again, move your left hand to your right elbow as in the first phase and throw the coin into your arm where it vanishes. Reproduce the coin again with your right hand at your left elbow.

With the two coins, perform the "Ten-Count Coin Trick" as a finale.

Vanishing Coins with Rings

Here's a standalone effect that allows you to use two rings to vanish a coin. The rings are not finger rings, but larger metal rings that you can purchase at a hardware store. The effect is rather offbeat but is quite visual and convincing.

Effect

By using a set of rings and a playing card, you cause a coin to vanish and then reappear.

Materials and Requirements

Two prepared rings.

A coin.

A couple sheets of matching paper.

A playing card.

To Prepare the Ring

Glue.

Scissors.

Secret

One of the rings is gimmicked.

Preparation

You'll need to purchase two rings from the hardware store. These are metal and about two inches in diameter. If you can find plastic rings or even thick bracelets, these should work as well. The rings must have a certain thickness to them to accommodate the gimmicking.

You'll need a piece of paper to perform the trick on. You can choose whether to use plain white paper or to go with color.

Take one of the rings, place it onto one of the sheets of paper, and using a pen or pencil, trace the inner diameter of the ring onto the paper, as in Figure 7.70.

Figure 7.70
Trace the inner diameter of the ring onto the paper.

Cut out the circle. It's best if you cut the circle a bit larger at first. This way, you can trim it down as you need to for the next step.

Compare the size of the circle against the ring. You want a circle that can be glued to the bottom of the ring but completely fills the bottom of the ring without sticking out on the sides.

When the circle is the right size, glue the circle onto the bottom of the ring, as in Figure 7.71.

Figure 7.71
Trim the paper circle to the correct size and glue it to the bottom of the ring.

You now have a gimmicked and normal ring, as shown in Figure 7.72.

Figure 7.72
You now have a gimmicked and normal ring.

Performing the Trick

Lay both rings on the paper and slide part of the playing card underneath the normal ring so you can see it within the ring. Also, set down the coin. The playing card is a "convincer" that suggests that both rings are normal. You can see this in Figure 7.73.

Figure 7.73
Lay the rings on the paper with the playing card.

Take the playing card and lay it on top of the gimmicked ring, as shown in Figure 7.74.

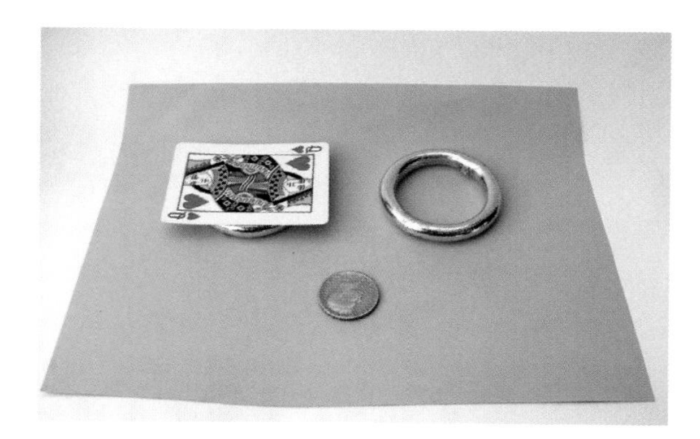

Figure 7.74
Stack the playing card on top of the gimmicked ring.

Take the normal ring and lay it on top of the playing card, as in Figure 7.75.

Figure 7.75
Place the normal ring on top of the playing card.

Pick up the stack of rings as a single unit and place it on top of the coin, as in Figure 7.76.

Figure 7.76
Place the stack of rings on top of the coin.

Remove the top ring and playing card to show that the coin has vanished. Actually, the coin is resting under the paper that's inside of the gimmicked ring as in Figure 7.77.

To show that the coin is gone, you can lift the entire paper to show that there's nothing underneath.

Figure 7.77
The coin has vanished.

To bring back the coin, stack the playing card on top of the gimmicked ring and the normal ring on top of the playing card, and then remove the entire stack. The coin reappears.

Lay out the rings and playing card in the original configuration when you started.

You can create something of a kit to carry around and perform this effect. Just fold the paper so that it contains and carries the rings and playing card. When you're ready to perform, unfold the paper and set up your rings and playing card. After you're done, fold everything up so you can perform the trick again.

Another thought. If you perform this trick on the side or lid of a case that you use to carry magic props, you can fill the gimmicked ring with material that matches that of your case. This will remove the need to perform the effect on a sheet of paper and will create a trick that will seem more impromptu and convincing.

The only thing that you'll have to be careful of is making sure that the gimmicked ring doesn't rest on a crease in the paper, which will expose the method.

The Washer on the String

Effect

You are able to remove a washer, nut, or finger ring that's threaded onto a string that is held by a spectator at both ends. The washer mysteriously passes through the string.

Secret

You use two washers and switch them.

Materials and Requirements

Two identical washers, nuts, or rings.

A length of string—two feet is good.

A handkerchief that you can't see through.

Preparation

None.

Performing the Trick

Finger palm a washer in your right hand and hand the string and other washer to a spectator. Allow the spectator to examine the washer and string. Ask the spectator to thread the washer onto the string and hold the ends of the string so there's no way for you to remove the washer, as in Figure 7.78.

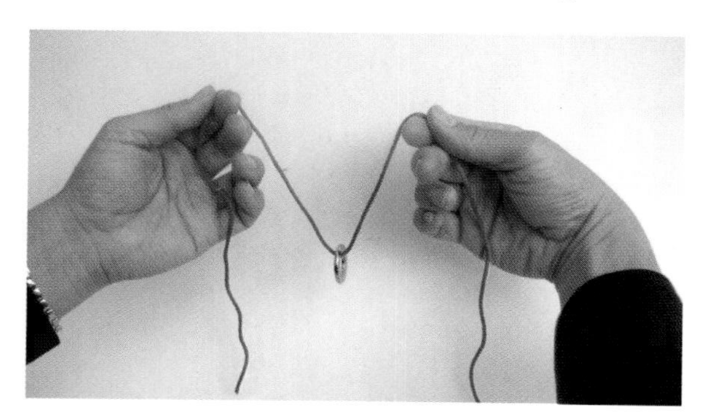

Figure 7.78
The spectator threads the washer onto the string.

Bring out the handkerchief and drape it over the washer and string. As in Figure 7.79.

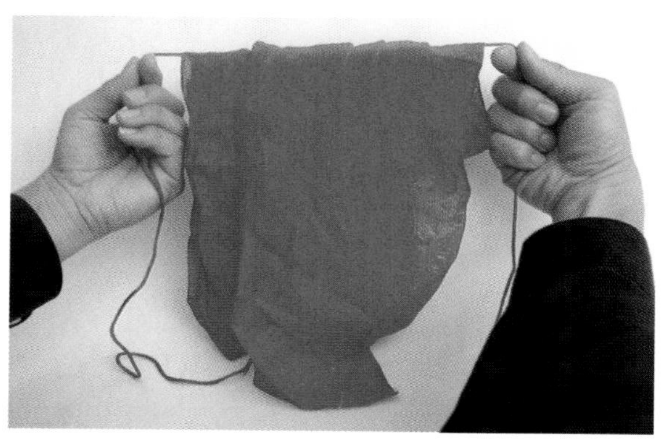

Figure 7.79
Drape the handkerchief over the washer and string.

Place both hands under the handkerchief and while there, out of the spectator's sight, do the following.

Bring out your second washer from finger palm and position it to the right of the threaded washer. Taking some slack in the string, push it through the hole in the washer, as in Figure 7.80. Don't hesitate to ask the spectator to give you some slack should you need it.

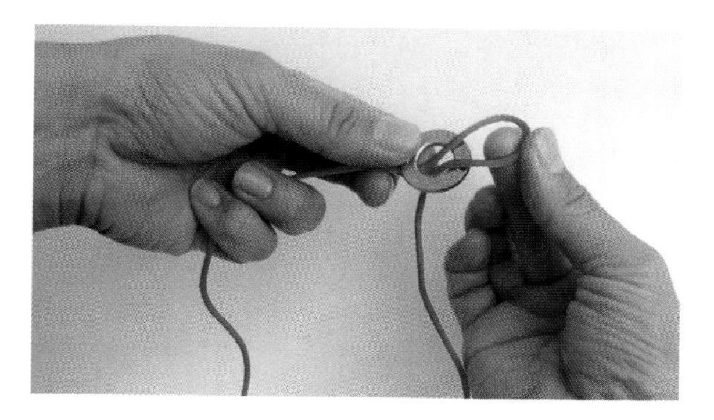

Figure 7.80
Under cover of the handkerchief, push the string through the hole of the second washer. We've removed the handkerchief for demonstration purposes.

After pushing the string through the washer, loop it around the edges of the washer so it stays in place, as shown in Figure 7.81.

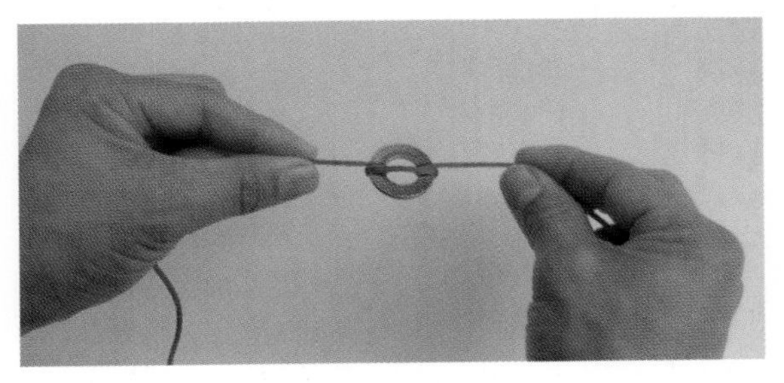

Figure 7.81
Loop the string around the second washer so that it stays in place.

Remove your hands from under the handkerchief. Grab the handkerchief with your left hand and grab the first (threaded) washer that's on the string. Slide the handkerchief and washer to your left until it's near the spectator's hand. This unveils the second washer. The spectator will think that this is the original washer.

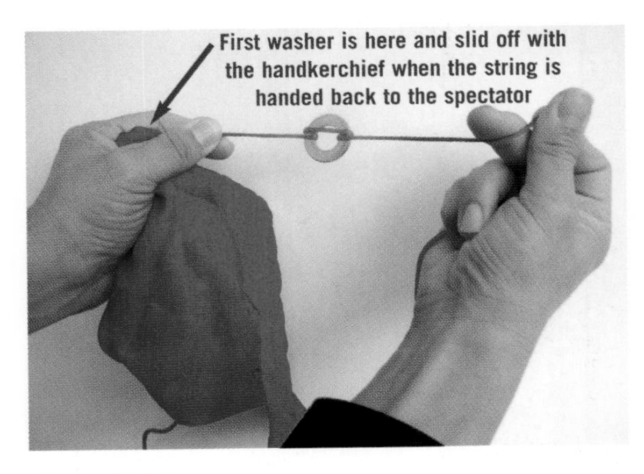

Figure 7.82
First washer is hidden in the handkerchief.

The second washer acts as misdirection. The spectator will see the second washer, which takes his attention away from the handkerchief and the first washer.

Encourage the spectator to examine the second washer. Grab the string and first washer through the handkerchief and grab the other end of the string with your free hand and lift the string up so the spectator gets a better look. This encourages the spectator to release his grip on the string.

Remember to keep some tension on the string to ensure that the second washer doesn't release itself prematurely.

Bring the string back down and slide the handkerchief to the end of the string. Ask the spectator to again grip both ends of the string. The washer is now bundled up in the handkerchief and off of the string. Place the handkerchief with the washer into your pocket.

Using both of your hands, release the loop around the washer and allow the loop to fall through the hole, thus releasing the washer. The washer has somehow passed through the string. The spectator may examine everything.

Next Steps

After this chapter on close-up magic, you can move onto Chapter 8, which discusses mentalism and mind reading. Here you'll learn tricks that make it look as if you have predicted the future and can see into peoples' minds.

8

Mind Reading and Future Predictions

IF YOU WANT TO READ SPECTATORS' MINDS, predict the future, envision images that you can't possibly see and more, a particular branch of magic known as mentalism may be for you. Mentalism is magic that is literally a state of mind.

While mentalism is clearly entertainment, it's a specific branch of magic where the line between tricks and real feats, psychic phenomena, for example, are often blurred. Some mentalism pros preface their performances by saying "if I could really predict the future, it might look something like this." While others simply present their feats as if they were real.

Because of the pseudo scientific nature of the tricks, they are often referred to as "experiments." And many performers act as if they have no idea how these supernatural powers work.

Predictions Versus Card Tricks

CARD TRICKS WHERE YOU PREDICT a card that a spectator will select are essentially mentalism effects. While playing cards are sometimes used in mentalism routines, audiences often assume that a certain level of trickery, sleight-of-hand and such, makes the card magic happen.

Mentalism, on the other hand, strives to create the impression that a phenomena is actually happening, without the benefit of trickery. Thus, while playing cards are a natural for mentalism, they are only sometimes employed.

In this chapter, I've provided a variety of effects that will allow you to perform different feats of mentalism. These effects rely on various techniques and secrets and provide an introduction to this particular branch of magic.

Visualize the Tricks

Before you read the description and secret of each trick, you may find it helpful to view the performances of them on the accompanying DVD. Actually, I recommend that you watch the performance of each trick before you read its description and secret. For this, please refer to the section on "Mentalism" that demonstrates the effects.

Your Destiny

This effect makes use of a concept that magicians call the "multiple out." It means that no matter what a spectator chooses, you already have the means to magically reveal it, just as if you had known all along or actually predicted the event. This one requires some preparation ahead of time, but it's a great trick.

Effect

The spectator is shown three envelopes, labeled "A," "B," and "C," and a poker chip. The spectator is asked to lay the poker chip on any one of the three envelopes. You show that you knew all along which envelope the spectator was going to pick.

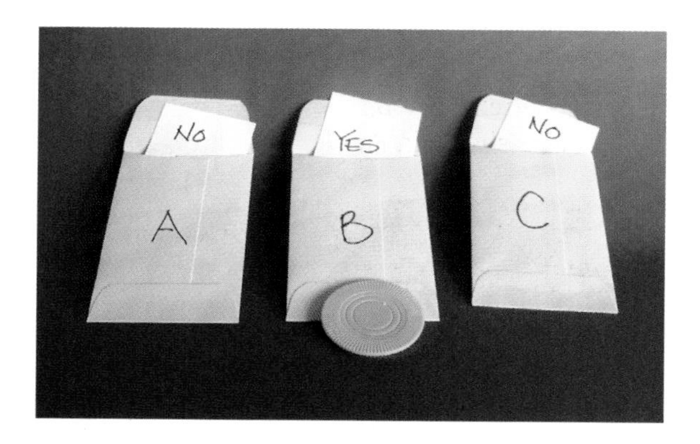

Materials and Requirements

Three envelopes (these can be full size for a stand-up show or tiny cash envelopes for close-up).

Three cards of heavy stock that fit inside the envelopes.

A poker chip.

A marking pen or Sharpie.

Secret

Regardless of the envelope that the spectator selects, you have the means to show that you predicted each choice. You have "multiple outs" or outcomes.

Preparation

This one can be done at any time, but you have to prepare the envelopes as follows:

Turn the envelopes so their address sides are face down on a table.

Write the letters: "A," "B," and "C," one on each envelope, as shown in Figure 8.1.

Figure 8.1
Designate and write "A," "B," and "C" on the envelopes.

Take the three cards and write "yes" on one, and "no" on the remaining two, as shown in Figure 8.2.

Figure 8.2
One card has "yes" written on it and two cards have "no."

Turn over the envelopes so you're looking at the front, address sides. Write "yes" on the "C" envelope and "no" on the other two, as shown in Figure 8.3.

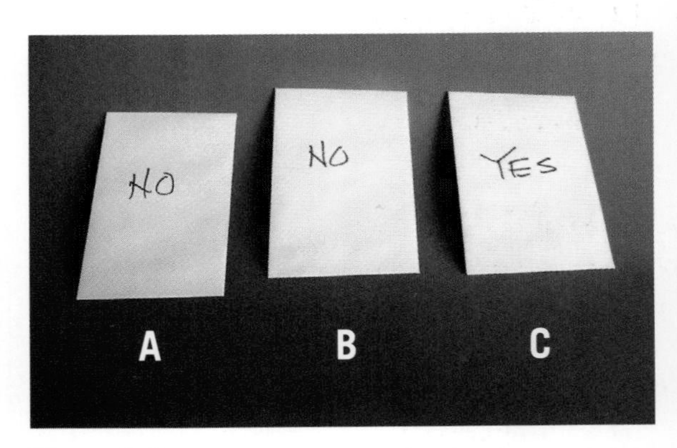

Figure 8.3
Two envelopes have "no" written on them and "C" has "yes."

Take the poker chip and if it's a light colored chip, say white or yellow, write the letter "A" on it using the marker. If it's a dark colored chip, you may want to put a sticker on the chip or glue a small label to the chip that has the letter "A" on it, as shown in Figure 8.4.

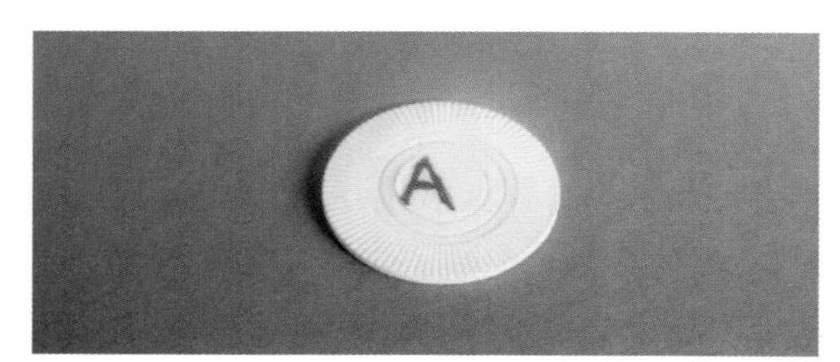

Figure 8.4
Write "A" on the poker chip.

Place the "yes" card into the envelope labeled "B." Place the "no" cards into envelopes "A" and "C." If the spectator selects "B," your outcome or "out" is the "yes" card inside.

Lay out the envelopes onto the table so their "A," "B," and "C" labels are displayed. Figure 8.5 shows the setup to begin the trick.

The "C" envelope is the only one that says "yes" on its underside. If the spectator selects "C," your out is the backside that says "yes."

Turn the poker chip so its label is face down on the table and can't be seen. If the spectator selects envelope "A," you'll turn over the chip to imply that you knew he would select "A."

Figure 8.5
The starting layout for the trick.

Performing the Trick

Ask the spectator to slide the poker chip, "the marker," and rest it in front of one of the three envelopes.

If the spectator places the poker chip onto envelope "A," turn over the chip to show that you knew he was going to select envelope "A," as shown in Figure 8.6.

Figure 8.6
Turn over the chip when the spectator selects envelope "A."

If the spectator places the poker chip onto envelope "B," pull the cards out of the envelopes and show that envelope "B" was the only envelope with a card that said "yes." Be careful not to show the fronts of the envelope or the other side of the poker chip. This is shown in Figure 8.7.

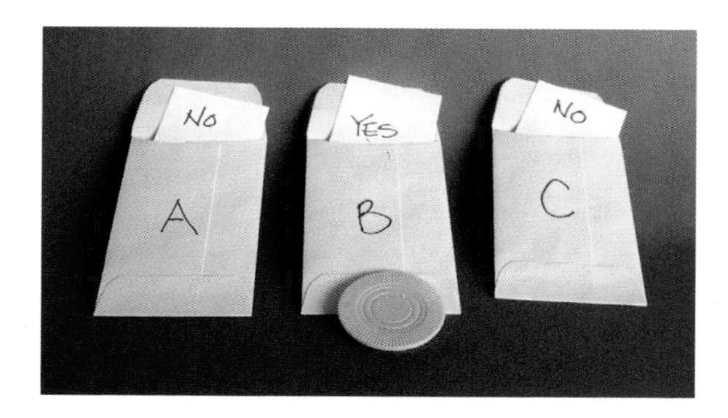

Figure 8.7
When the spectator selects envelope "B," show the cards in the envelopes.

If the spectator places the poker chip onto envelope "C," turn over the three envelopes and show that envelope "C" is the only envelope that has "yes" written on it. Be careful not to flash the cards inside the envelopes or show the other side of the poker chip. This is shown in Figure 8.8.

Figure 8.8
When the spectator selects envelope "C," show the backs of the envelopes.

When performing tricks with multiple outs, it's important to collect the materials and get them out of the reach of spectators who may want to examine them. Because spectators can't examine the materials without discovering the secret to the trick, be sure to have another trick on hand so you can immediately move onto something else.

Your Destiny 2

This effect relies on a well-known and interactive force. All you need are six objects and a spectator to perform the magic.

Effect

Six objects are gathered and placed onto a table. You bring out a piece of paper and write a prediction on it. You fold the paper and set it aside where it's in plain view. You and the spectator take turns eliminating objects. When there is a single object on the table, you show that you predicted that it would be left. The name of the object is written on the paper.

Materials and Requirements

Six small objects.

A piece of paper.

A pen.

Secret

This effect is based on a simple force that leaves an object of your choosing. When you write your prediction, you're free to select any one of the six items. But when you and your spectator take turns eliminating objects, you each select pairs of objects and allow the other person to pick the object to eliminate.

When you pick your pair of objects, you always make sure that the pair does not contain your predicted object. And if the spectator creates a pair that includes your object, you always eliminate the other object. In this manner, you force your predicted, written-down object.

Preparation

Gather and lay out the six objects.

Performing the Effect

Take out the paper and select and write down one of the objects that you see on the table. Do not allow spectators to see this. Fold the paper and leave it out in plain view. For purposes here, we'll assume that you selected the Rubik's Cube, as shown in Figure 8.9.

Figure 8.9

For demonstration purposes, we'll assume that you selected the Rubik's Cube.

Pick two items, remember to not include your predicted item, the Rubik's Cube, in the pair, and allow the spectator to eliminate one, as shown in Figure 8.10.

Place the chosen object to the side where it's out of play.

When the spectator selects two items, pick one to eliminate. If the spectator happens to include your object in a pair, eliminate the other object, as shown in Figure 8.11.

Figure 8.10

Pick two objects and make sure that this pair doesn't include your force item. Let the spectator select one for elimination.

Figure 8.11

If the spectator includes your object in his pair, select the non-force item to be eliminated.

Continue taking turns selecting and pairing items and eliminating an object.

At the end, you'll come down to the final two objects and it will be your turn to eliminate an object. One will be your predicted object. Just eliminate the other object.

Ask the spectator to pick up the prediction and read it. It will correctly predict the remaining object, as shown in Figure 8.12.

Figure 8.12
The force results in your predicted object remaining on the table.

You can perform this force with different items, but the procedure varies with the number of objects. If there's an odd number of objects, the spectator has to go first to find a pair and let you eliminate an object. If there are an even number of objects, as was described here, you must go first to establish a pair of objects and allow the spectator to eliminate one.

Got Your Number

This prediction effect works with the animated cartoon that is found in the back of the book. The animation is in the lower-right corner of each page. As you flip through the pages of the Appendix from back to front, you can watch the resulting animation, which reveals the number four, as shown in Figure 8.13.

Figure 8.13
The animation in the lower-right corner of the back of this book works with this effect and reveals the number four.

This effect is based on a force. Beyond the revelation here, you can use this method to force any other number, letters, or even words or pictures. But to support the animated revelation in the back of this book, we'll show you how to force the number four, which is what's revealed in the cartoon.

Effect

You bring out ten small squares of paper and write the numbers one through ten on each. The papers are placed into a cup or paper bag and then the spectator is asked to toss the slips onto the table. All of the slips that land blank-side up are removed.

The rest of the slips are placed back into the bag and tossed out again. This is repeated until there is one slip remaining. You show that you predicted this outcome by flipping through the back pages of this book, which shows an animation that reveals the same number.

Materials

Ten small slips of paper. These don't have to be perfect and you can cut them out using scissors.

A bag or cup.

This book.

You'll need to perform this one on a table.

Secret

One of the slips has the same number written on both sides, which forces the number four.

Preparation

Beforehand, write the number four on one of the slips of paper. This is the force number that is revealed in the animation. Set this one with the number side down. It's best to set this slip aside so you can readily identify it or you can place a small mark on the paper so you can immediately recognize it.

Performing the Trick

Bring out the slips of paper and lay them down on the table. Be careful not to show that one slip already has a number written on it, as shown in Figure 8.14.

Figure 8.14
Lay the slips on the table.

Write the number "one" on the first slip and place the slip into the cup or paper bag. Repeat with the numbers "two" and "three." When you get to "four," find the slip that already has a "four" on its underside, and write "four" on the top side and place this into the cup or bag, as shown in Figure 8.15. Be careful not to allow a spectator to see the number four that is already written on the bottom of the slip.

Figure 8.15
Write "four" on the top side of the slip that already has "four" written on the bottom side.

Continue writing numbers on the remaining slips.

Ask the spectator to toss the slips onto the table and remove the slips that come up blank and set them aside, as shown in Figure 8.16.

Figure 8.16
Remove the slips that turn up blank.

Place the remaining slips back into the cup or bag and toss again.

After all of the slips have been removed, the number four will remain. Note this to the spectator and then put all the slips away and out of play.

Go to the back page of this book and flip through the pages from back to front and show the spectator the animation in the lower-right corner of the screen. The animation will conclude with the revelation of the number four. This is shown in Figure 8.17.

Figure 8.17
Flipping through the pages to view the animated revelation.

The Clock

Appendix C presents an interactive trick. By following along with the pages, you can perform a feat of mentalism. We explain the trick here, as there's no background information provided with Appendix C.

Figure 8.18
This trick relies on the instructions and revelation in Appendix C.

Materials

Appendix C of this book.

Secret

The revelation for this effect is the number six, which is forced through the instructions in Appendix C. The secret is based on the way that numbers are displayed on a clock face.

When you create pairs of numbers by matching those that are directly across from each other on the clock's face, and subtract the smaller number from the larger number, you always end up with the number six.

Performing the Trick

Turn to the first page of Appendix C. Read aloud the instructions and allow your spectators to refer to the images. When you reach the last page of the appendix, it will reveal the number six, which is the force number that your spectators will reach, which is shown in Figure 8.19.

Figure 8.19
Despite the seemingly free choices, readers will each conclude with the number six.

One Ahead

The name of this trick is also the secret. Known as "one ahead," this method allows you to seem to be writing one answer, when you're secretly writing another. As you'll see, it's a sneaky and effective way to secretly read answers and appear to predict the future.

Effect

You write down a prediction on a slip of paper, fold it, and set it aside in plain view. You ask a spectator to name his favorite color, which he does.

You write a prediction on a second slip of paper, fold it, and set it aside in plain view. You ask a second spectator for the amount of change in his pocket.

You write a prediction on a third piece of paper, fold it, and set it aside. You ask a third spectator to select a playing card. When you open the slips of paper, you show spectators that your predictions were correct.

Materials

Three slips of paper.

A pen.

A deck of cards.

Secret

You force the playing card and write this as your first prediction on the slip. When you're apparently writing your second prediction, you're actually writing down the first spectator's answer. In this manner, you are "one ahead." At the end, you can show the correct answers, but you have actually written them out of order.

Skills

The ability to perform a cross-cut force with playing cards, as explained in Chapter 4.

Preparation

Place the four of clubs on the bottom of the deck in preparation for the cross-cut force.

Performing the Trick

Take out your slips of paper, look at a spectator, and then write down "four of clubs" on the first slip. Fold the paper and set it aside. It's important that your spectators don't see what you're writing.

Ask the spectator to name her favorite color. For dramatic purposes, you can nod to acknowledge that you may have gotten it right. Let's say that the spectator said "purple." The situation is shown in Figure 8.20.

Figure 8.20
You've written your force card on the slip before the spectator calls out a color.

Take the second slip of paper and write "purple." Again, make sure that the spectators don't see what you are writing. Fold the slip of paper and set it aside. Ask a second spectator for the amount of change in his pocket. Let's say that the spectator says 32 cents. Again, nod in acknowledgment. The current situation is shown in Figure 8.21.

Figure 8.21
You now have two predictions on the table—your known, forced card prediction and the spectator's color.

> **Reality Check**
>
> The spectators think you've written down a color before asking a spectator to name a color. They also think that you've written down a quantity of money before asking the second spectator. In reality, you've written down your force card to fulfill the prediction of the third outcome, and used this opportunity to write a "prediction" after you've heard it. This is the "one-ahead" principle in use as you start out "one ahead," by knowing one of the outcomes.

Take the third slip of paper and write "32 cents." Bring out the deck of cards and prepare for the cross-cut force. As you're waiting after the spectator cuts the cards, you can go over the first two predictions. Ask the spectator to lift the top section of cards where they "cut," and name the card. You can simply leave the cards out, as they are ungimmicked and may be examined.

Pick up your slips of paper and hold them in your hands. Mention to spectators that there were three free choices that you predicted on the slips of paper. At this point, you can shake the slips of paper in your hand so spectators can't tell which slip is which prediction.

Open the slips of paper and show that you had successfully predicted each outcome, as shown in Figure 8.22.

Figure 8.22
You've correctly predicted the outcomes.

> **Note**
>
> It's important that your writing somewhat resembles the prediction. Thus, in the example above, when you're writing "32 cents," you may have to pretend to be writing a few extra strokes so it looks as if you are writing the name of a playing card. It's attention to details such as this that make the magic trick successful.

The crux of the "one-ahead" principle lies in the fact that you already know the outcome of one prediction. As a result, once you thoroughly understand the principle, you can use any force for your third "prediction," which includes the number force used in the "animated prediction," or "Your Destiny 2," and the "Envelope Switch," which is explained next.

At the Movies—The Envelope Switch

Here's another trick that's based on a well known forcing technique. And yes, the name is the method. With this technique, a spectator can make a completely free choice by choosing from a number of slips that are in a transparent plastic bag. But in putting the slip into an envelope for "safe keeping," you make a diabolical switch to a note of your liking.

Effect

You show a clear plastic kitchen bag that has a number of pieces of paper in it. You ask several spectators to reach into the bag, pull out a slip of paper, and to read what's on it aloud. Each reads the name of a different movie, refolds the slip, and places it back into the bag. After three spectators have selected and read the contents of a slip, state that the next slip will be the one used in the experiment.

Invite another spectator to select a slip and then put it into an envelope for "safe keeping." The envelope is handed back to the spectator to hold. You direct everyone's attention to an envelope that has been in plain view and state that your favorite movie is written inside. You open the envelope and show that you had written "Star Wars."

The spectator who has been holding the envelope with the chosen slip is asked to open the envelope and read the slip of paper inside. It's found to say "Star Wars." You have effectively used a force, both literally and figuratively.

Materials

A gallon-size, clear plastic kitchen bag (Ziploc brand bags are good for this).

21 slips of paper about two-inches square.

An 8.5" x 11" Manila mailing envelope.

20 small 2.25" x 3.5" envelopes, sometimes called "coin" or "pay" envelopes.

A pen.

A sheet of paper.

A rubber band.

Secret

The spectator's choice of paper from the plastic bag is completely free. You switch the slip of paper when you place it into the envelope.

Preparation

Write the name of a different movie on each slip of paper. If you will be forcing "Star Wars," use the name of other movies on these slips.

Fold all of the slips and place them into the plastic bag, as shown in Figure 8.23.

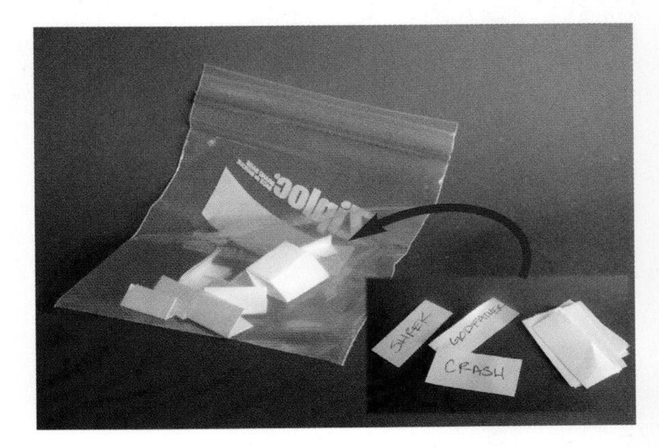

Figure 8.23
All of the slips in the bag have the name of a different movie written on them.

Take one of the small envelopes and cut off its sealing flap, as shown in Figure 8.24.

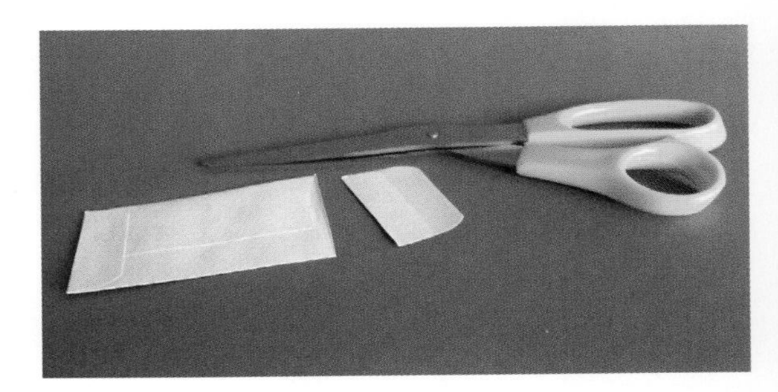

Figure 8.24
Cut off the flap to an envelope.

Write "Star Wars" on a slip of paper and insert it into a different envelope, as shown in Figure 8.25.

Figure 8.25
Write "Star Wars" on a slip of paper and insert it into an envelope.

Figure 8.26
The envelope with the prediction is placed on top of the stack and the envelope that is missing its flap is placed on top of the envelope with the prediction.

Stack the rest of the small envelopes so they're facing the same direction, with the "address" side to the back and the "seam" side facing you.

Place the envelope that contains the "Star Wars" slip of paper on top of the stack and then place the gimmicked envelope that is missing its flap on top of the entire stack, as shown in Figure 8.26. Note how the sealing flap of the second envelope is allowed to extend up.

Wrap the entire bundle of envelopes with a rubber band. Be sure that the envelope with the prediction and the gimmicked envelope that is missing its flap are aligned on top. It should look as if you simply have a stack of envelopes that are held together by a rubber band, as shown in Figure 8.27. The flap over the top envelope is really attached to the second envelope.

Figure 8.27
From all appearances, you have a stack of envelopes that are held together by a rubber band.

Performing the Trick

Bring out the plastic bag and tell spectators that you have written the names of movies on the slips of paper. Invite different audience members to reach into the bag, pull out a slip of paper at random, and read the title of the movie aloud. You'll hear spectators reading different titles to prove that the bag contains different movie titles.

Ask each audience member to fold and return his or her slips of paper to the bag.

Bring out your stack of envelopes and state that the next selection by an audience member will be the title that will be used for this experiment.

An audience member reaches into the bag and pulls out a slip. You put down the bag and ask the audience member for the slip so you can put it into an envelope for "safe keeping."

Holding the envelopes in a downward and open position, place the slip of paper into the front envelope, the one without the flap. Be careful not to separate the front envelope from the one directly behind it as it lacks a flap. You want spectators to see you openly placing the chosen slip into the front envelope. This is shown in Figure 8.28.

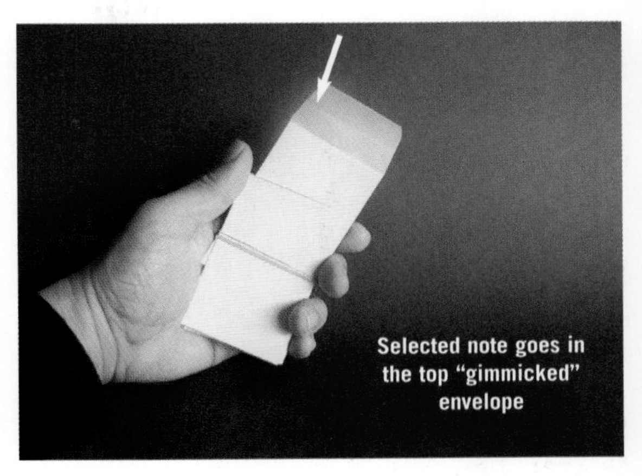

Figure 8.28
Place the chosen slip of paper into the front (gimmicked) envelope.

Grab the top flap, which is really the flap to the second envelope with the predetermined slip of paper. Bring the stack of envelopes up and pull out the second envelope, as shown in Figure 8.29.

Figure 8.29
As you bring the stack up, pull out the second envelope. Note that this is your view. The spectators should not see the envelope being pulled out from this angle.

As you pull out the second envelope, look the spectator in the eye and state, "I'm going to have you hold this envelope for safe keeping." This is misdirection that should keep the spectator's mind off the moment when you switched the envelopes.

Hand the envelope (the second envelope) to the spectator. Put away your stack of envelopes to hide the evidence.

Bring out your prediction envelope that has been sitting in plain view the entire time. Pull out its sheet of paper and read "Star Wars" aloud. Freely show the sheet of paper front and back so spectators can read it for themselves.

Ask the spectator to open his envelope and read the name of the movie, which will be "Star Wars."

While I've explained this envelope switch to support this effect, you can use this force for any number of applications. For example, in the previous effect, "One Ahead," you could easily combine this technique with the "one-ahead" principle to force the last parameter, which can be anything that you want.

Elephants in Denmark

This is an interactive trick that's presented in Appendix D. By following along with the instructions on those pages, you perform a feat of mentalism with your spectator. The title says it all as this is the revelation for the trick. Of course, you won't want to refer to the trick by this name to your spectators.

Materials

Appendix D of this book.

Secret

The revelation is the country of Denmark and Elephants, which is forced through the instructions in Appendix D.

The secret is based on a phenomenon that is associated with the number nine and its multiples. If you take the two digits of any multiple of nine (18, 27, 36, etc.) and add them together, you always get nine (18: 1+8=9, 27: 2+7=9, 36: 3+6=9, etc.).

By forcing nine and then subtracting five, this forces the number four, which is then associated with the letters of the alphabet. This, in turn, forces the letter "D," which is the fourth letter in the alphabet. The first country that most people will think of that starts with the letter "D" is Denmark (you may occasionally get a "Dominican Republic," which blows the trick).

The spectator is then asked to take the second letter and think of an animal that begins with this letter. Of course, most spectators will think of "elephant," although you may occasionally get an "eel."

Performing the Trick

Turn to the first page of Appendix D and read aloud the instructions and allow your spectators to refer to the graphic images. When you reach the last page of Appendix D, it will reveal "Denmark" and display a picture of an elephant, which is the conclusion that your spectators will reach, as shown in Figure 8.30.

Figure 8.30
Spectators will end up with "Elephants in Denmark."

It's best to perform this trick for groups of people. This way, you'll have several people who will come up with the right answer. If you only perform for one person, there's a chance that they will make a mistake and come up with the wrong answer, or come up with "Dominican Republic" instead of Denmark, or "eel" instead of "elephant." By performing this trick for several spectators, you're guaranteed that someone will arrive at the correct answer.

Mind Reader

In this effect, we teach you how to make a gimmicked deck of cards that lets you see the exact card that a spectator has glimpsed. You'll have to dedicate a deck to this trick, but it can serve you well.

Effect

You hold a deck so it's facing a spectator and slowly riffle through it and ask a spectator to say, "stop." The spectator is free to choose any card and can see different cards going by. When the spectator says, "stop," you close the deck and then proceed to read the spectator's mind and tell him his selected card.

Secret

You've prepared a special mind reading deck that tells you the exact card that a spectator has selected. The deck takes some work, but you'll be able to use it in a fantastic feat of mind reading.

Materials

A special deck of cards. You'll have to dedicate a single deck to this effect.

Glue (a glue stick works well for this).

A pen.

Preparation

Divide the cards into two packs. It's best to shuffle the deck so they are thoroughly mixed. Take half of the deck and try to make sure that there's a good blend of red and black cards and representation of face cards. If you need to, shift cards around so the two halves have even distributions of high and low and red and black cards.

Take half of the deck and using a paper cutter, trim about a sixteenth of an inch off the top of the short end of each card, as shown in Figure 8.31.

Figure 8.31
Trim approximately a sixteenth of an inch from the top of each playing card.

Trim the top corners of each card with scissors to round them so they look like normal cards, as in Figure 8.32.

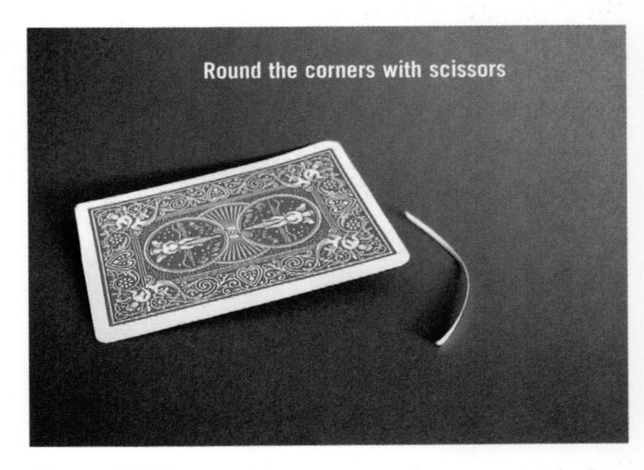

Figure 8.32
Trim the corners to round them and make them appear to be normal cards.

Many craft stores sell a corner cutter that can round the corners for you. These can be purchased for a few dollars. You have just created a "short card," a card that is shorter than a standard card.

Take a regular card and a short card. Place the short card in front of the long card with the cards facing forward, as shown in Figure 8.33.

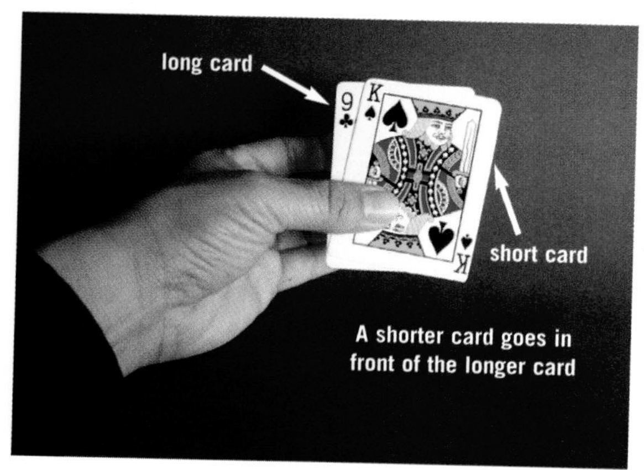

Figure 8.33
The shorter card goes in front of the longer card.

Line the cards at the bottom and glue the bottom portions together, as shown in Figure 8.34. You will be gluing together 26 pairs of cards.

Figure 8.34
Pair a long and short card and glue a strip at the bottom together.

Allow the cards to dry.

You'll have a deck of 26 pairs of cards that are glued at the bottom with a short card in the front.

Using the pen, take each pair and write the name of the long card on the top border of the back of the shorter card, as shown in Figure 8.35. You'll notice that we have simply referred to each card using two symbols, it's rank and suit. Also, we've written the code in both of the top corners.

Figure 8.35
Write the name of the long card on the top of the backside of the short card.

When you stack these gimmicked cards with the openings facing up, these cards behave in a peculiar way. Like a Svengali deck that we discussed in Chapter 6, the cards fall in pairs when you riffle them on the top with your finger, as shown in Figure 8.36.

Figure 8.36
The cards fall in pairs when you riffle the deck from the top.

Even better, when a spectator says, "stop" and is looking at a card, you can see the name of the card by looking at the back of the card in front of it, as shown in Figure 8.37.

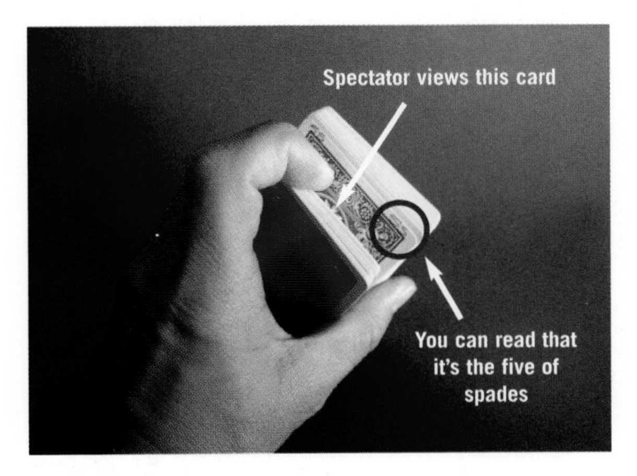

Figure 8.37
You can read the name of the card that the spectator is looking at.

The deck appears to be fairly normal. Like a Svengali deck, you can freely cut the deck and with some care, riffle shuffle it. Just be sure that you don't change the orientation of the cards, as was discussed in the section on stripper decks in Chapter 6.

Performing the Trick

Take out the gimmicked deck. Show spectators that the deck has different cards and mix it is as you wish. Because the cards are glued together, you can openly fan the deck and spread the cards.

Holding the deck in your left hand, riffle the top of the deck with your right index finger and ask your spectator to say, "stop."

When your spectator says, "stop," you glance at the back of the card in front of his card and secretly read the name of his selected card. Ask the spectator if he's "got it," and then close the deck and put it away.

From here, it's all acting. Pretend to try and read the spectator's mind and slowly begin to conjure up the image of their card, red or black, the suit, and finally, the name. Feel free to milk this for all it's worth.

It's in a Book

Mentalists often perform an effect known as a "book test." The mentalist says that she wants a spectator to think of an object and will try to read his mind. However, the mentalist will want a new thought, which will be easier to "read." And a good way to introduce a new thought into a spectator's mind is to have him randomly select a word out of a book.

Effect

You ask a spectator to think of any three digit number and then perform some math functions to arrive at a random number. You ask the spectator to open a book to the arrived at number and silently read the first word on the page and concentrate on it. After some thought, you are able to tell the spectator the word he is thinking of.

Secret

The final "random" number is forced through some clever mathematical operations. The end result will always be the number "1089." You tell the spectator to use the first three digits, which will be "108." Of course, beforehand, you've already checked the word on page 108 in the book and memorized it.

Materials

A pad of paper for the spectator to use.

A pencil or pen.

A book with at least 200 pages.

Preparation

Beforehand, open the book to page 108 and remember the first word on the page.

Performing the Trick

Provide the spectator with the pencil and pad.

Ask the spectator to write down any three-digit number.

Ask the spectator to reverse the three-digit number and write this one underneath the first.

Ask the spectator to subtract the smaller number from the larger number. If the new number is a two digit number, tell the spectator to add a zero to the beginning of the number. For example, if the spectator ends up with 89, by adding a zero to the beginning of this number, they will use the number 089.

Ask the spectator to reverse the new number and add it to the original number. The spectator is to remember this number. You do not look at the number, but it will be "1089." The mathematical operations are shown with an example in Figure 8.38.

1. Write down any three-digit number.

2. Reverse the number and write the second number directly below the first number.

3. Subtract the smaller number from the larger one. If this new number only has two digits, place a zero in front of it.
For example, 87 becomes 087.

4. Reverse this new number and add it the previous number.

5. Remember the number.

Example:
Spectator's three-digit number: 983
Reversed: 389
983 - 389 = 594

Reverse 594 to make 495

594 + 495 = 1089

Figure 8.38
The math operations that support the book test.

You ask the spectator to use the first three digits, which will represent a page number. Of course, the page number will be "108," and the spectator does not know that you are ahead of the game.

Ask the spectator to open the book to his page number and silently read the first word on the page. Ask the spectator to concentrate on the word.

After taking a moment to pause and appear to be concentrating, tell the spectator the word that he is thinking of, the word that you memorized earlier.

Note

You can perform this trick with any book, just as long as you look at page 108 ahead of time. This trick can be particularly baffling if you are able to secretly glance page 108 in a book that the spectator owns or is carrying with him.

Since this method always forces the number "1089," you can use this procedure as the basis for any trick where you wish to reveal the number "1089," which can refer to a quantity of objects, an ID number, finances, an address, the last four digits in a phone number, and lots more.

You can improve the impact of this trick by offering two books for spectators to choose from. If you are only working with two books, it's not that hard to remember two words on page 108 of both books. Just be sure to remember which word is associated with which book.

You can also perform what is called a "magician's choice." Here, you place the two books on the table and ask a spectator to point to one. If the spectator points to the book that you want him to use, you simply hand the book to him. If the spectator points at the "wrong" book, the one that you don't want to use, simply take it away and say, "alright, we'll use this one."

Notice that you simply ask a spectator to point to a book. You never ask him to point to the book that he wants to use. It's subtle wording that makes all of the difference and allows the force to work.

©istockphoto.com/Kati Neudert

Call the Psychic

This is a variation on a classic that uses a telephone. In the old days, this effect was often performed at parties where everyone could gather around a telephone. These days, with cell phones, you can perform this miracle anywhere.

Effect

You bring out a deck of cards and allow a person to mix the cards and freely choose one. Or you can simply ask a spectator to name any card with the exception of the joker. The spectator names a card and you tell him about your amazing friend, "the psychic."

You bring out your cell phone and dial a number and ask for "the psychic." You give the phone to your friend or turn on the speaker, and the psychic names the spectator's card.

Secret

You have a friend who's working with you and understands a secret code.

Materials

A cell phone.

A deck of cards (optional).

Preparation

You and your friend, who you will call, have to understand the code, which is explained in the next section.

Performing the Trick

Allow a spectator to mix the deck and freely choose a card, or have them name any card in the deck except the joker. You can say that the joker is a card that is hard for your friend to get an impression from.

Once the card has been established, bring out your cell phone and call your friend. When he answers, you say, "hello, may I please speak with the psychic?"

Your friend slowly says "lower" and then "upper."

If the spectator's selected card is a six or below, say, "yes, the psychic," after your friend says "lower." If the spectator's selected card is a seven or above, just allow your friend to say "upper" and continue to the next step.

If you've said "yes" after the psychic says "lower," your friend knows that the selected card is a "six" or below, and then goes through the lower range of cards as follows: "ace," "two," "three," "four," "five," and "six." When your friend says the value of the card, say, "Hello, this is (your name)."

If you've allowed your friend to say "upper" without you saying anything, your friend now knows that the selected card is a "seven" or higher. He mentions each card in order starting with "seven" as follows: "seven," "eight," "nine," "ten," "jack," "queen," and "king." When your friend says the value of the card, say, "Hello, this is (your name)."

In either case, your friend now knows the value of the card.

Your friend then names the suits, "clubs," "hearts," "spades," and "diamonds." After your friend mentions the suit of the spectator's card, say, "yes, I have a friend here who has selected a card, can I put you on speaker phone?"

Provide a moment to pause as if to allow your friend to answer "yes," and then turn on the speaker phone.

When the speaker phone is ready, say "Please name the card."

Your friend names the card and hangs up. If you like, you can say that your psychic friend is rather abrupt.

It's best to use your own phone to perform this trick and not a spectator's phone. If the number you are calling gets placed into a spectator's phone, he may try and call it later and get a hold of your friend.

Next Steps

In the next chapter, you'll be delving into the world of stand-up and stage magic—magic that you can use to entertain bigger crowds.

9

Stand-up and Stage Magic

MOST OF THE TRICKS IN THIS BOOK are appropriate for close-up situations where you're performing for up to ten people at a time. But a big part of magic is performing for bigger groups and in theaters. When you're on a stage, people in the back row will have a hard time seeing your close-up tricks so you must perform effects that "play big" and can be seen from a distance.

Stand-up magic is quite different from close-up magic. For one thing, you usually have more control over your environment. In most stand-up situations, people are seated and watching you and you're likely standing on a stage or to one side of a room so everyone can watch you. As a result, you can worry less about people watching you from behind or from the sides who may see your secrets.

In fact, stand-up magic often features effects that can only be performed on a stage in a controlled environment. And these effects count on the fact that no one is standing behind you or to your sides. Also, the more controlled environments of stand-up magic often use music.

Types of Stage Magic

FOR STAGE AND STAND-UP WORK, there's a branch of magic that's known as "manipulation," which is essentially large-scale sleight-of-hand that's used to vanish and produce handkerchiefs, balls, and other objects, as well as live birds.

Another branch of stage magic is that of illusions. Technically, an illusion is any trick that you perform, but the term is often used to describe large-scale magic tricks that include sawing a person in half, levitating a person, making a car disappear, producing a helicopter, and more. Illusions typically require assistants and a dedicated stage crew, and to be successful at this type of magic, you must have expertise in lighting, choreography, sound, and more. Performing large-scale illusions is an art unto itself.

While manipulation and illusions are technically "stage magic," we're going to focus on "stand-up" magic—tricks that you can perform while standing in front of groups of people. In stand-up magic, you'll typically have a case from which you can pull props, and you can generally count on your audience being seated in formal or semiformal situations.

Some of the tricks will require that you make props, but you should be able to find and purchase all of the necessary raw materials from an office or craft store. Some may require you to print items from your computer and none are expensive.

The Missing Pip

This is a card trick for the stage. While it's a card trick in the vein of "pick a card," its revelation can be seen from a distance. The trick requires that you make a prop that will appear to be a large playing card.

Effect

You have a spectator select a card, show it to the audience, and place the card back into the deck and mix it up so the card is lost. You try to find the spectator's card and fail. After several attempts, you turn around a board that depicts a large playing card. You reveal the card, which is the "five of diamonds."

The spectator tells you that this is the wrong card. After looking surprised, you ask the spectator to name her card. She says that it was the "four of diamonds."

Without missing a beat, you say, "no problem," and proceed to flick the middle pip of the large card off to display a "four of diamonds," as shown in Figure 9.1. Actually, it's a card with four pips and the number five on it.

Figure 9.1
Your five of diamonds turns into the spectator's card, the four of diamonds.

Secret

You force the four of diamonds and have a specially made jumbo card that allows you to make the surprising change from a five of diamonds to a five of diamonds that has only four pips.

Skills

Ability to force a card.

Ability to false cut and false shuffle a deck (optional).

Preparation

You have to construct a large-scale card that has a removable center pip.

Materials

A heavy, poster-sized board. For this, foam core boards are probably best. You can also make a smaller 8.5" x 11" board.

Two large number 5s and five diamonds that are printed from a color inkjet printer. (If you only have access to a black-and-white printer, you can use spades or clubs, but you will have to locate clipart for the pips.) While it's best to use clipart for the diamonds, you can draw your own on your computer or even by hand directly onto the board if you need to.

You'll want a sheet of thin but sturdy cardboard to act as the back of the "removable" diamond.

Glue (a glue stick works fine).

Rubber cement.

Scissors.

If you want a professional-looking prop, you'll want to create a back that looks like that of a giant playing card. For this, you can purchase gift wrap in a pattern and color that looks something like a card back. I recommend that you use a bright weave or geometrical pattern. Also, a large paisley pattern can work well.

Another option, party supply stores often have giant cards for sale that are used as decorations for casino nights. You might be able to locate one of these and alter it for use in this trick.

Preparation

To help create your own card, we have provided artwork in Appendix F that you can photocopy and then glue onto your giant card. Note that we have written up this trick to work with a "five of diamonds" and Appendix F provides artwork for a "five of clubs."

The black card is a lot easier to copy and work with. Just alter the following instructions to work with a "five of clubs" anywhere the "five of diamonds" is mentioned.

If you are accustomed to working with a PC, here are instructions for creating and printing your own card elements. Print large 5s and five diamonds from your computer's printer to fit your giant card. To do this, you can use a word processing or desktop publishing program to manipulate a basic font. To be seen in proportion to your giant card, start out with a point size of 300 and adjust the size until it looks right for your card. You'll need to print two 5s, as shown in Figure 9.2. If you like, you can also use stick-on letters that you purchase from a craft store.

Figure 9.2
Print two 5s.

In the same manner, locate clipart or a font that has a diamond. Some programs install a font that features symbols, which often have diamonds, hearts, clubs, and spades that you can use. You can size these just as you would a normal font.

Otherwise, you'll have to search clipart libraries in the programs that you have or search the internet for a suitable diamond, heart, club, or spade symbol to use. You can also create squares in a graphic draw or paint program and rotate them 45 degrees to create diamonds. You'll need to print five matching "pips," as shown in Figure 9.3.

Figure 9.3
Print five diamonds.

If you want to finish the back of your giant playing card, take your wrapping paper and cut a piece to fit the back. You can choose to fit the wrapping paper to the edges of the "card" so it fills the entire back or cut the gift wrap paper to allow for a white border. Glue your wrapping paper to the back of your giant playing card, as shown in Figure 9.4.

The cleanest way to secure the number and pips to your giant card is to cut them out and glue them to your poster board. At this stage, cut out and glue the two 5s and four of the pips in an arrangement that mimics that of an actual five of diamonds playing card.

Figure 9.4
Glue wrapping paper to the back of your "card" to finish it and make it look like a jumbo card.

For the final pip that goes in the center, glue the big pip onto a cardboard backing and then cut it out. Place a dab of rubber cement in the back of the pip and stick it onto the board in the center. You'll have to experiment with the amount of glue so you can "flick" the pip off if you choose to. If you like, you can also simply remove the pip.

On stage, turn the card around so spectators can't see the five of diamonds. It looks best if you display the back of the card on an easel of some kind.

Prepare a deck of cards so you can force the four of diamonds using your favorite technique.

Performing the Trick

Bring out a deck of cards and perform a false shuffle and cut, if you like.

Allow a spectator to select a card and force the four of diamonds. Use your favorite force here.

Because of the stand-up conditions, it may be best to use the Hindu or Cut Deeper forces that don't require a table as the cross-cut force does. When interacting with audiences, in this case, I think it's best to go out into the crowd and have someone select a card. You can find instructions on performing these forces in Chapter 4.

Ask the spectator to return the card to the deck and then mix up the deck.

Tell the spectators that you will find their card and then show the top card, which isn't their card. Act as if you don't know what their card is, and then direct their attention to the giant card on stage. Tell the audience that this is a giant prediction card. Turn the card around and show them the five of diamonds.

When the audience tells you that it's not their selected card, ask for the name of the card. When the spectator or another audience member says "four of diamonds," react by saying, "hey, at least I was close." This usually gets a laugh.

It's all acting now. Ask the audience if they really wanted you to find the "four of diamonds" and tell them that you're prepared to deliver. Walk up to your giant card or hold it in your left hand. It's more dramatic to flick the center pip off the board to show that the card is now the "four of diamonds." You can also simply grab the center pip and remove it and transform the giant card.

I love performing these types of revelations where it appears that I have lost a card. And when I later reveal the card in a surprising way, it adds to the impact.

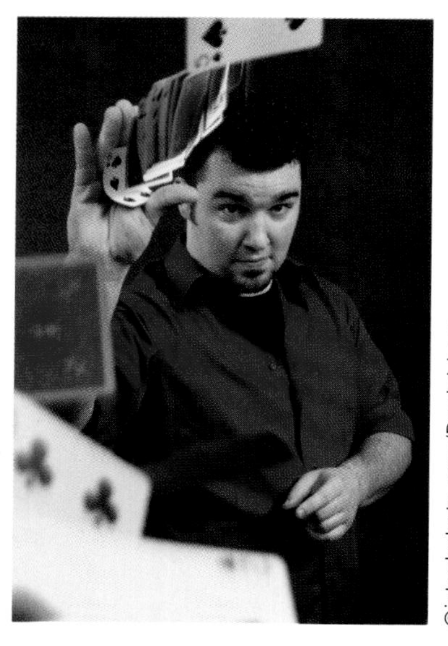

©istockphoto.com/Daniel Lemay

The Magical Album

This is a well-known effect that is often performed at kid shows and is also known as the "coloring book." While this prop is available at any magic store, the advantage of making your own is that you can customize the pictures to suit any theme that you wish. We explain here an album that has pictures of playing cards, but the pages can turn into a stamp collection, a fairy tale book with the right magic spell, and more.

Effect

You show your audience a photo album or book and thumb through its pages to show that it's empty, as shown in Figure 9.5.

After a magical gesture or asking audience members to wave their hands towards the book in a magical fashion, the book fills with pictures, as shown in Figure 9.6.

It's a visual effect that works with lots of themes.

Figure 9.5
The book is empty.

Figure 9.6
The book is now full of pictures.

Secret

The book is like a Svengali card deck that you learned about in Chapter 6. The pages alternate between long and short so when you thumb through the pages of the book in one direction, the pages appear to be empty. And when you thumb through the pages of the book the other way, the pages are filled with whatever pictures you've placed on them.

Preparation

This one is all about constructing the prop.

We offer three different types of books that you can gimmick:

1. You can gimmick a standard photo album, the kind with plastic cover sheets that you pull back to insert photos. Since you'll be cutting the outside edges of the page, make sure that if you do cut this edge, you won't lose the plastic cover sheet.

2. You can gimmick a standard composition or blank hardback book. You'll simply glue your pictures into its pages.

3. You can print pages on a computer or glue pictures onto individual pages, alter the pages as we'll discuss here, and then have them bound with staples or binder clips, or even have the book professionally bound. We won't specifically offer a procedure for creating this book, but it should be clear from the instructions that we provide.

What you'll need to do is cut every other page so it's shorter along the outside edge that you will "thumb." The first page is shorter, the second page is normal, the third page is shorter, and so on. An eighth of an inch should be fine, depending on the firmness of the pages.

With every other page cut "short," you can insert or glue pictures onto every other pair of pages: 2 and 3, 6 and 7, 10 and 11, and so on. Be sure to test the book before you take the effort to glue in all of the pictures throughout the entire book. The general order of the pages is shown in Figure 9.7.

Figure 9.7
Cut every other page so it's slightly shorter. Images are face to face.

When you flip through the book from back to front, the pages will appear blank. And when you flip through the book from front to back, the pages will appear to be full of pictures. Because of the short and long nature of the pages, the pages fall into pairs, and depending on the direction that you thumb the pages, only certain pages display.

Performing the Trick

Present the book to the audience and by flipping through the book from back to front, show them that the book is empty. Close the book.

Depending on your theme, for example, you can simply say that you will use magic to fill up the book with life and pictures. Open the book and flip it from front to back to show that the book has magically filled with pictures.

You can purchase a professionally made magic coloring book for less than twenty dollars. However, you're stuck with the theme of an empty coloring book that magically fills with color. By building your own prop, you can apply an endless number of themes that will work better with your stage persona.

©istockphoto.com/Mirela Schenk

Ropes Through Body

Here's an interactive trick that allows you to work with three volunteers from the audience. It's like a poor man's sawing a person in half, but you're using ropes. If you like, you can also "saw" yourself in half.

Effect

You bring three spectators up on stage. You display two long ropes and hand two ends to one volunteer and the other two to a second volunteer. The center of the ropes go behind the third volunteer.

You take a single end from each volunteer who is holding a rope and tie a simple knot in the rope and then hand the ends back to each volunteer. You tell the volunteers with the rope that when you count to three, you want them to pull on the ropes, seemingly trapping the spectator in between.

You count to three and the volunteers pull, and the ropes seem to pass right through the body of the third volunteer. Two normal ropes are now in front of the spectator and the ends are still held by the pullers. At this point, the volunteers can pull on the ropes all they want, they are normal.

If you like, you can stand in-between two volunteers and have the ropes pass through your body. Also, you can perform this one on yourself, which is how we demonstrate the effect in the pictures.

Secret

You prepare the ropes so that they are never actually around the body of the third volunteer as they may appear.

Materials

Two lengths of rope, about six feet in length each. It's best to use a soft rope because your volunteers will be grabbing and pulling it. Clothesline is fine.

Some thread (preferably white).

Preparation

Fold each rope in half, as shown in Figure 9.8.

Secure the bends of both ropes together with some thread. You'll want to experiment with the thread as you'll want enough thread to connect the ropes together so they will be secure and won't break prematurely. But you'll also want the threads to break easily when spectators pull on them. This will depend on the strength of the thread that you are using.

Figure 9.8
Fold each rope in half.

To start, try wrapping the ropes together with two loops of thread and then tying the thread. This is shown in Figure 9.9.

Figure 9.9
Secure the ropes together at their center points using a couple wraps of thread. Note that for demonstration purposes, we have used dark thread. You will want to use white thread that won't contrast and show.

When you hold the ropes at the center and cover the thread with your hands, to your audience, it looks as if you're holding two ropes, as shown in Figure 9.10.

Figure 9.10
Hold the two ropes with the "join" in your hand. It will look to spectators as if you are holding two separate ropes. Note that this is the exposed view that you would see if you opened your hand.

Performing the Trick

Bring up your volunteers onto stage and take out the ropes. Hold the ropes with your hand covering the thread.

Invite one volunteer to stand in the middle and give two ends of the rope to each volunteer on the side. You may want to have the volunteers move forward a bit so they can't see the thread behind the third volunteer.

Make sure that the thread is hidden behind the third volunteer. You may also need to ask the other volunteers to hold the rope more tightly and adjust the position of their ropes. You can also ask the third volunteer in the center to hold the ropes behind him to steady it.

If you're performing the trick on yourself, place the two ropes behind your back, with the "join" in the middle of your back, as shown in Figure 9.11.

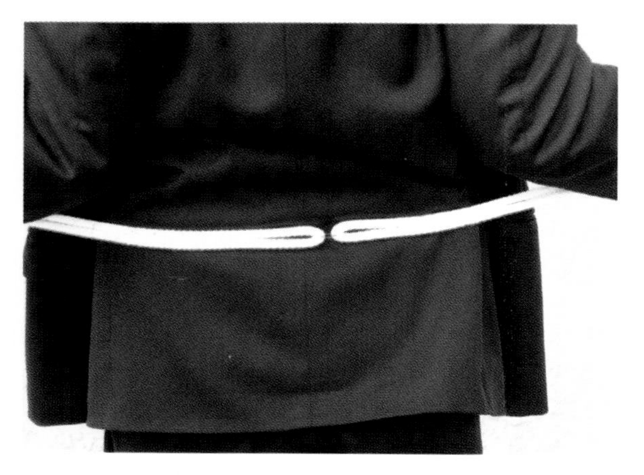

Figure 9.11
Place the two ropes behind your back with the "join" in the middle of your back.

Move to the front of the third volunteer and take an end from each of the other volunteers and tie a simple knot, as shown in Figure 9.12.

If you're performing the trick on yourself, tie the simple knot in front of yourself using one rope from each side. It's the first half of a standard square knot.

Give the ends back to each spectator.

Tell the spectators that when you count to three, you want them to pull hard on both ropes. Count to three and when the spectators pull on the ropes, the thread will break and the ropes will appear to pass through the third volunteer.

Figure 9.12
Tie a simple knot.

If you're performing the trick on yourself, just hold the ropes in your hands and pull, as shown in Figure 9.13.

Figure 9.13
The ropes appear to pass through your body.

Thank your volunteers and encourage the audience to give them a round of applause as they go back to their seats.

Cut and Restored Rope

Here's a classic of magic. You cut a rope and then restore it. We'll teach you the traditional method. It's going to take some work to learn, but once you've got it, you can always perform this one.

Effect

You cut a rope, tie the two pieces together, and then remove the knot and restore the rope.

Secret

The trick involves a move that makes it look as if you are cutting the center of the rope when you are actually cutting the rope near its end. The rope will be slightly shorter at the end of the routine, but audience members won't notice this.

Materials

A rope, about four or five feet in length. Softer rope is better as it's easier on your hands. Clothesline is fine.

A pair of scissors that can easily cut the rope.

Preparation

None.

If the rope is frayed at the ends, simply cut off the ends just before your show so the rope isn't frayed, which will interfere with the trick.

Performing the Trick

Bring out the rope and snap the rope to show that it's solid. If you like, you can hand the rope out for examination.

Grab the leftmost end of the rope with your left hand. Grab the middle of the rope with your right hand. Grab it as shown in Figure 9.14. Notice that the right hand is palm up and the rope is grasped between the first finger and thumb.

Figure 9.14
Grab the middle of the rope with your right hand.

Bring the center of the rope towards the left hand, and using the right first and second fingers, clip a portion of the rope just below the left hand, as shown in Figure 9.15.

Figure 9.15
Use the first and second fingers of the right hand to clip a portion of the rope just below the left hand.

Pull the portion of the rope held by the right first and second fingers, point "A," underneath the portion held by the first finger and thumb, as shown in Figure 9.16.

Figure 9.16
Pull the first and second fingers under the original center of the rope.

Bring the new "center" of the rope up where it's held by the left hand, as shown in Figure 9.17. Note that the point where the ropes crossover remains behind the hand and hidden from spectators.

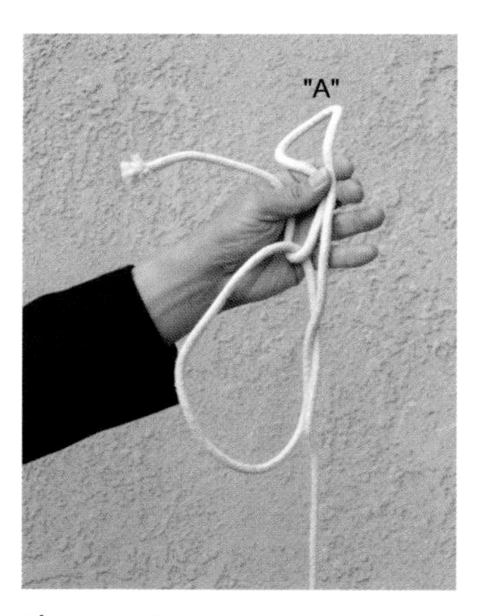

Figure 9.17
The new "center" of the rope.

> **Reality Check**
>
> It appears that you've grabbed the center of the rope and brought it up to be held by the left hand. What you have really done is switched a portion of the rope near the end for the center of the rope, and are now holding this portion in your left hand. The spectators will think you are holding and cutting the center of the rope.

While we've explained the move here in steps, it should be performed in one fluid motion. This is the crucial part of the trick that you'll want to practice until it's second nature. Figure 9.18 shows what it looks like from the spectators' standpoint.

Figure 9.18
It looks as if you've grabbed the center of the rope to prepare to cut it.

Place your scissors into the loop, as shown in Figure 9.19.

Figure 9.19
Insert your scissors into the loop.

Cut the "center" of rope that you are holding in your left hand, as shown in Figure 9.20.

Figure 9.20
Cut the fake center of the rope.

Allow the newly cut right-most end to drop. Figure 9.21 shows your view.

Figure 9.21
Drop the newly cut end.

Reality Check

The two ropes look to be about the same size, but because of the way that you are holding the ropes, one rope is actually far longer, about 90% the length of the original rope, and the second rope is far shorter. The shorter rope is looped with the longer rope and each rope is double backed. The ends at the top of your hand are actually the ends of the shorter rope. The ends dangling down are the ends of the longer rope.

Figure 9.22 shows the audience's view.

Figure 9.22
It looks like you have two ropes of nearly the same size (audience view).

Be careful that you don't show spectators that the ropes are double-backed and looped together. Tell the spectators that you will now restore the rope.

Taking your free right hand, begin to tie the short ends at the top of your left hand into a knot. You'll want to tie a simple square knot, as shown in Figure 9.23. Don't tie this knot too tightly as you'll need to slide it off shortly.

Figure 9.23
Tie the short ends into a square knot.

With the knot tied, you can release one side of the rope. It will look as if you have two ropes that are relatively the same length tied together. State that you have successfully restored the rope. With the knot in the middle, this should get some laughs and protests from your audience.

Reality Check

It looks as if you have two ropes of similar length tied together, but you actually have one long rope and a short rope. The short rope is tied to itself to look like a knot.

Tell the audience that you didn't realize that they wouldn't like the knot. Hold the rope at one end with your left hand and using your right hand, wind the rope around your left hand, as shown in Figure 9.24.

Figure 9.24
Begin to loosely wrap the rope around your left hand.

When your right hand reaches the "knot"—the short rope that is tied to appear as a knot—simply grab the knot with your right hand and slide it off of the rope as you're winding the rope around your left hand, as shown in Figure 9.25.

When you're done winding the rope, the knot or short rope will come off in your hand.

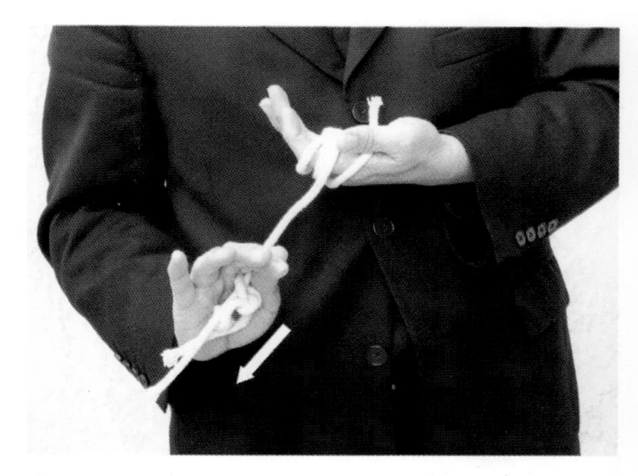

Figure 9.25
Grab the short rope, which appears to be a knot, with your right hand. As you're winding the rope around your left hand, secretly slide the knot off of the rope.

Hint:

Depending on how you set up the rope, you may have a knot with long ends that may be hard to grab and hide in your hand. If you have such long ends, you can simply cut the ends off before proceeding to wind the rope around your other hand and removing the knot.

Immediately reach into your pocket with your right hand to find the scissors. Grab the scissors and leave the short end of the rope in your pocket. Wave the scissors over your left hand (safety first, hold the blades in your right hand as you would when walking with the scissors). Put down the scissors or replace them in your pocket.

Using your right hand, slowly unwind the rope off of your left hand to show that it has completely restored itself.

Dollar Bill to Book

Here's an effect where a borrowed dollar bill vanishes and reappears in an impossible location. While this effect uses a dollar bill that can't actually be seen from a distance, your interaction with the crowd allows it to work for a crowd. I've seen similar effects play to entire theaters.

Effect

You borrow a dollar bill, write down its serial number, place it into an envelope for "safe keeping" and allow an audience member to hold the envelope. When the audience member is asked to open the envelope and look inside, it's found to contain an I.O.U. A book that's been sitting out in plain sight is picked up by a spectator and when the front cover is opened, there's a dollar bill and the serial number matches that which is written on the envelope.

The borrowed bill appears to have vanished from its envelope to be found in a book, and the serial number acts as proof that it's the same bill.

Secret

You switch the borrowed bill with the I.O.U. slip and use the serial number of a bill that you have already placed into the book. This trick relies on the same method as that used in "The Envelope Switch," which was explained in Chapter 8.

Materials

A dollar bill—you want a bill in medium condition that is not brand new, but is not old and rumpled. This way, your bill will most likely resemble the one that you borrow.

A book.

10 small 2.25" x 3.5" envelopes, sometimes called "coin" or "pay" envelopes.

A pen.

A small slip of paper approximately the size of a dollar bill.

Preparation

Take one of the small envelopes and cut off its sealing flap, as shown in Figure 9.26.

Figure 9.26
Cut the flap off of an envelope.

Write "I.O.U." on the slip of paper and insert it into an undoctored envelope. On the bottom of the envelope on the "seam" side, write down the serial number of the dollar bill, as shown in Figure 9.27.

Figure 9.27
Write "I.O.U." on the slip of paper and insert it into an undoctored envelope. Note the serial number of the bill written on the envelope.

Stack the rest of the small envelopes so they're facing the same direction, with the "address" side to the back and the "seam" side facing you.

Place the envelope that contains the "I.O.U." note and has the serial number written on it on top of the stack. Then place the gimmicked envelope that is missing its flap on top of the entire stack, as shown in Figure 9.28.

Wrap the entire bundle of envelopes with a rubber band. Be sure that the envelope with the prediction and the gimmicked envelope that is missing its flap are aligned on top. It should look as if you simply have a stack of envelopes that are held together by a rubber band, as shown in Figure 9.29.

Place the dollar bill in the front of the book.

Figure 9.28
The envelope with the "I.O.U." and the serial number is placed on top of the stack and the envelope that is missing its flap is placed on top of the envelope with the prediction.

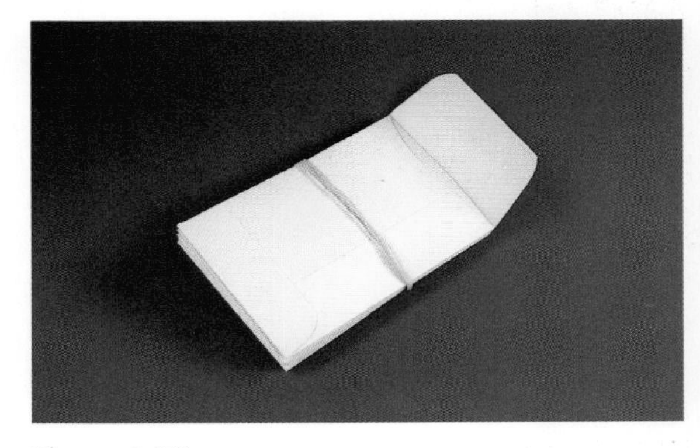

Figure 9.29
From all appearances, you have a stack of envelopes that are held together by a rubber band.

Performing the Trick

Borrow a dollar bill from someone in the audience.

Take hold of the stack of envelopes and with the bill and envelopes facing you, write down the serial number of the bill on the envelope using the pen. Do not let spectators see what you are writing. At this point, you can pretty much write just about any serial number you want as you'll soon be switching the envelopes.

After you "write" down the serial number, casually display the envelope to the audience, but don't give them time to read the serial number and possibly memorize it.

Holding the envelopes in a downward and open position, fold the borrowed dollar and place it into the front envelope, the one without the flap. Be careful not to separate the front envelope from the one directly behind it as it lacks a flap. You want spectators to see you openly placing the borrowed dollar into the front envelope. This is shown in Figure 9.30.

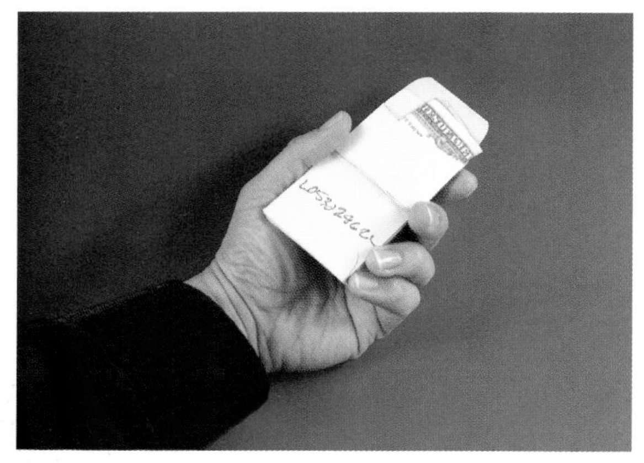

Figure 9.30
Place the folded bill into the front (gimmicked) envelope.

Grab the top flap, which is really the flap to the second envelope with the predetermined slip of paper. Bring the stack of envelopes up and as you bring the stack up, pull out the second envelope, as shown in Figure 9.31.

As you pull out the second envelope, look the spectator in the eye and state "I'm going to have you hold this envelope for safe keeping." This is misdirection that should keep spectators' minds off of the moment when you switched the envelopes.

Figure 9.31
As you bring the stack up, pull out the second envelope. Note that this is your view. The spectators should not see the envelope being pulled out from this angle.

Hand the envelope (the second envelope) to the spectator. Put away your stack of envelopes to hide the evidence. Be careful not to allow your audience to see the serial numbers that you wrote on the top envelope.

Bring out the book and hand it to the spectator who originally lent you the bill to hold. Ask the spectator not to open the book.

State that the magic has happened and ask the first spectator to open his envelope and take out the bill. The spectator will find the note stating "I.O.U." Ask the spectator to show the note to the audience.

Ask the second spectator to open the cover of the book and display what's inside. The spectator will find the dollar bill.

Ask the first spectator to read off the serial number that's written on the envelope and invite the second spectator to read along. Verify with the second spectator that the serial number on the bill matched that which was written on the envelope.

Full Count?

Here's another classic stage effect. While versions of this one are marketed commercially, we give you a way to make your own.

Effect

A card is shown with one diamond on one side and four diamonds on the other. The card is then shown to have three diamonds on one side and six diamonds on the other. Before spectators can catch their breaths, the card now has eight diamonds on one side.

Secret

The change occurs through the arrangements of the diamonds and the handling of the card—it's all based on where you hold the card when you present it, which allows you to show the different quantities of diamonds. A secret flap allows you to make the final change. For this, you'll have to make a special card.

Materials

Three sheets of black paper (we used standard-size sheets of black paper cut down to 7 inches by 11 inches).

White paper from which you can cut out diamonds.

Scissors.

Glue (gluesticks are fine).

Preparation

Take two sheets of black paper and fold them in half, as shown in Figure 9.32.

Figure 9.32
Fold the two sheets of black paper in half.

Glue together a half from each black sheet, as shown in Figure 9.33.

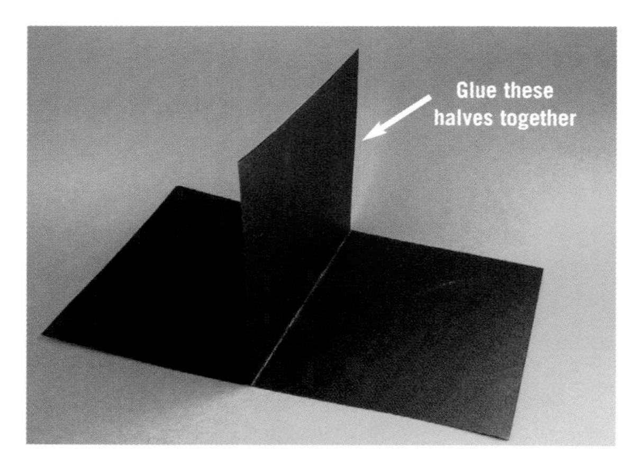

Figure 9.33
Glue half of each sheet together.

You now have two sheets of paper that are glued together in something of a "T" shape. Glue the other halves to the third sheet of black paper, as shown in Figure 9.34. You'll have a black card that has a flap that can be moved from side to side to show different configurations of diamonds.

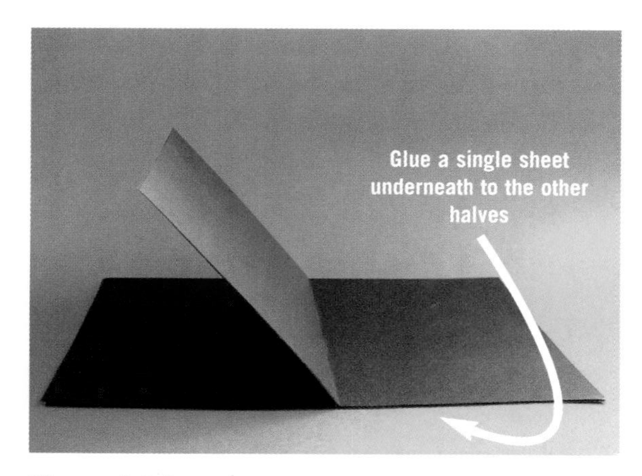

Figure 9.34
Glue the two sheets of paper to the remaining sheet of black paper.

Cut out 15 diamonds. Glue them in the following arrangements.

On the side without a flap, glue two diamonds, as shown in Figure 9.35. Note the location of the flap.

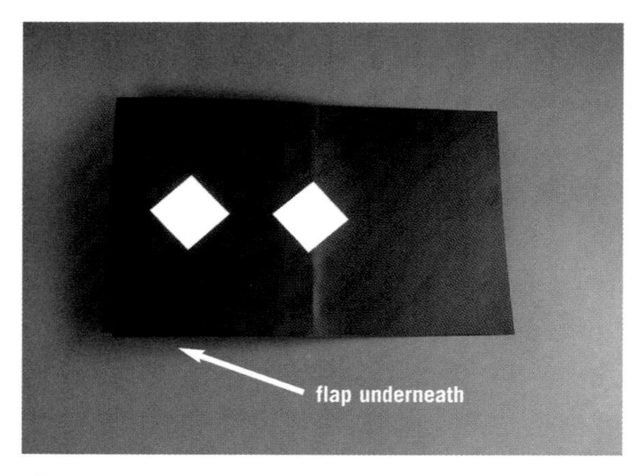

Figure 9.35
Glue two diamonds to side one.

On the side with the flap, glue five diamonds, as shown in Figure 9.36. Note the location of the flap. It's important that you orient the card correctly with the diamonds that you are gluing on.

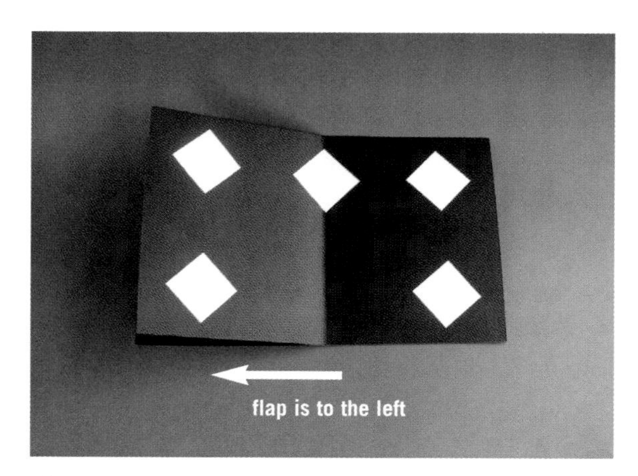

Figure 9.36
Glue five diamonds to side two.

On the same side, shift the flap to the other side and glue on eight diamonds, as shown in Figure 9.37.

Figure 9.37
Glue eight diamonds to "side three."

Using the sides with two and five diamonds, you can show different numbers of diamonds. For example, when you hold the card with your hand over one of the diamonds on the two-diamond face, it looks as if you have only one diamond, which is shown in Figure 9.38. When you hold the card with your hand over the blank space, it looks as if the card has three diamonds, which is shown in Figure 9.39.

Figure 9.38
Showing one diamond.

Figure 9.39
Showing three diamonds.

On the other side, in a similar manner, when you present the card and hold your hand over the middle diamond, it looks as if the card has four diamonds. And when you hold the card with the your hand over the blank spot, it looks as if the card has six dots. This will be demonstrated through the instructions that follow.

Performing the Trick

Bring out the card from your case with the two-diamond side showing and your right hand holding the bottom of the card, covering the bottom diamond. To the audience, it will look as if you have a card with a single diamond on one side, as shown in Figure 9.38.

Figure 9.40 shows your view and the inset shows the audience's view. From the back, grip your left hand over the middle diamond, with the back of your hand facing you, as shown in Figure 9.40, and turn over the card using your left arm as an axis.

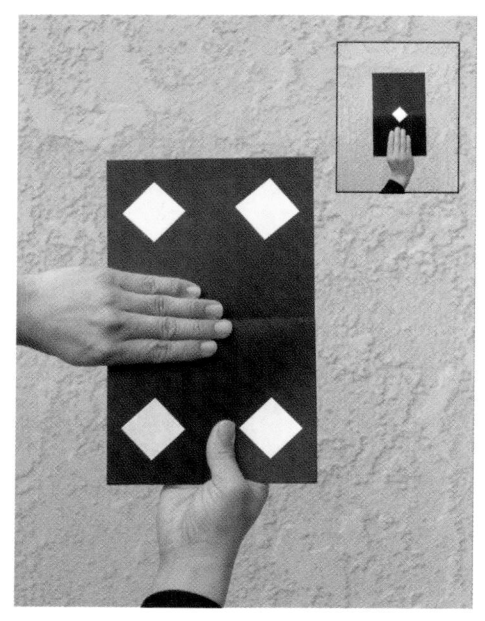

Figure 9.40

Bring out the card with one diamond showing. The main picture is your view and the inset is the audience's view.

Figure 9.41

The audience sees four diamonds. Grip with your right hand the blank spot on the bottom and turn the card over.

The audience now sees four diamonds on the other side, as shown in the inset of Figure 9.41. Reverse your right hand and grip the bottom of the card covering the blank spot, as shown in Figure 9.41. Note how the back of your right hand is towards you. Turn over the card using your right arm as an axis.

The audience will now see three diamonds, as shown in Figure 9.42. In the same manner as before, grip the card with your left hand and cover the blank spot. Note how the back of your left hand is towards you. This is shown in Figure 9.42.

Figure 9.42

The audience sees three diamonds. Grip the blank spot with your left hand and turn the card over.

The audience now sees six diamonds, as shown in Figure 9.43.

Figure 9.43
The audience sees six diamonds.

By gripping your right hand over the bottom dot and turning the card around with your right hand, you can show one diamond again and repeat the entire process.

After repeating the process, show that one side only has five dots and when you cover the middle dot, you can show the side as having four dots or six dots, as shown in Figure 9.44.

Turn over the card and show that the other side only has two dots, and when you cover the end dot or open space, you can show the side as having one dot or three dots, as shown in Figure 9.45.

Figure 9.44
You show how you can display the side as having four dots or six dots.

Figure 9.45
You show how you can display the side as having one dot or three dots.

While you're showing the side with two dots, secretly use your thumbs to flip the flap on the back of the card to reveal eight diamonds on your side, as shown in Figure 9.46.

Figure 9.46
Turn the flap to reveal the eight diamonds.

> **Hint**
>
> To help cover the shifting of the flap, you can ask spectators what a magician can do if the audience figures out how the trick is done. When spectators are looking at you, move the card up a bit and simultaneously shift the flap over. Tell them "this is what I do" and then turn over the card for the final revelation.

Use both hands to turn over the card to reveal the eight diamonds to the audience, as shown in Figure 9.47.

Figure 9.47
The audience sees eight diamonds.

> **Hint**
>
> If you're having trouble with the flap moving or not staying in place, you can try some velcro to keep the flap attached to the card.

Try to go for a smooth pace and practice until you don't have to think about your hand placement and the process feels completely natural to you.

This trick should be performed at some distance from your audience so they don't see the crease running down the center of the card. If you are standing closer to your audience where they can see the crease, you can try folding the entire card in half at the crease and inserting it into a big envelope. When you start the trick, you can bring out the folded card from its envelope and subtly justify the fact that there is a crease running down the middle – the card was just folded in half to fit inside the envelope.

The Three Rope Trick

Here's another great rope trick. A well known variation of this effect is called "The Professor's Nightmare" and is often performed by professional magicians. This is a "no cut" trick, which means that you can perform the trick over and over with the same ropes because you aren't cutting them. It's a classic stand-up trick that will serve you well and you can perform this one for close-up as well.

Effect

You bring out short, medium, and long ropes that become three ropes of equal length. At the end, the three ropes return to their original lengths: short, medium, and long.

Secret

You double-back the long rope, and when shown with the short rope, they appear to be two ropes that are the same length as the medium length rope.

Materials

Three ropes of the following lengths:

Short Rope: 20 inches.

Medium Rope: 38 inches.

Long Rope: 58 inches.

To prevent the ropes from unraveling, I often glue the ends and let them dry. This not only prevents the ropes from fraying and coming apart, it makes it easy to find the ends by touch. I highly recommend that you do this for your performance ropes.

Preparation

None.

Performing the Trick

Bring out the three ropes and clearly show that they are three ropes of different lengths. If you like, you can hand the ropes out for examination.

Place the short rope into your left hand, holding it near the end, as shown in Figure 9.48.

Figure 9.48
The short rope goes into your left hand.

Place the medium rope into your left hand, to the right of the short rope, holding it near the end, as shown in Figure 9.49.

Figure 9.49
The medium rope goes into your left hand.

Place the long rope into your left hand, to the right of the medium rope, holding it near the end, as shown in Figure 9.50.

Figure 9.50
The long rope goes into your left hand.

Bring the lower end of the short rope up and hold it in your left hand, to the right of the first three ropes, as shown in Figure 9.51.

Figure 9.51
Bring up the other end of the short rope and hold it in your left hand.

Bring the lower end of the medium rope up and hold it in your left hand, to the right of the first three ropes, as shown in Figure 9.52.

Figure 9.52
Bring the lower end of the medium rope up and hold it in your left hand, to the right of the first three ropes.

Bring the lower end of the long rope up and place it to the right of the ropes, as shown in Figure 9.53.

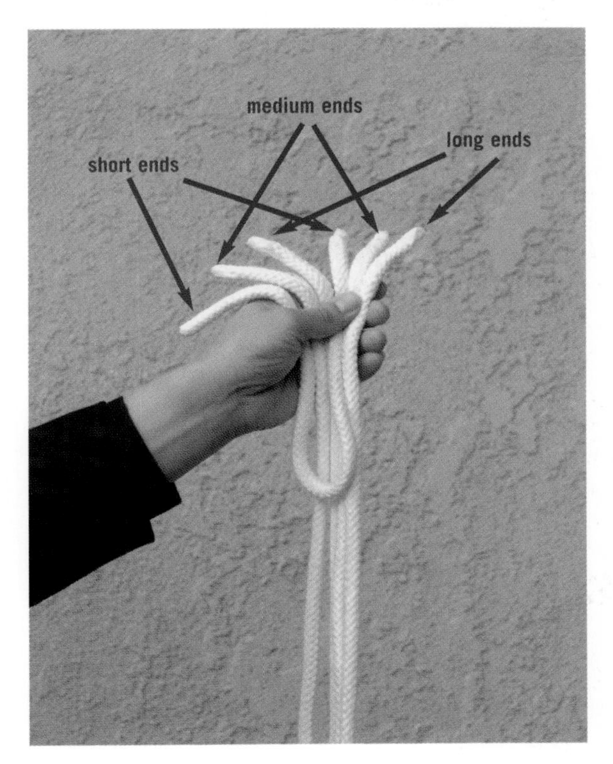

Figure 9.53
Bring up the remaining end of the long rope.

Here's the tricky part and the secret to the trick. Using your right hand, grab the two ends of the long rope and the rightmost end of the medium rope. This is shown in Figure 9.54.

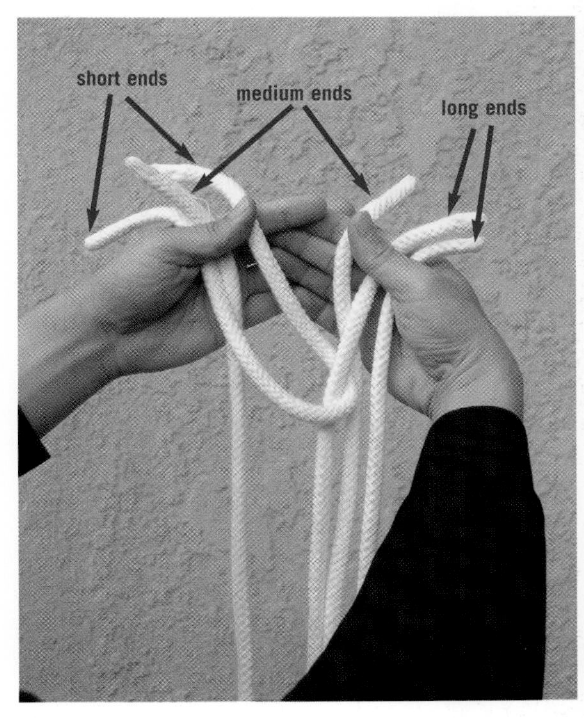

Figure 9.54
Arrange the ropes with the two short ends in your left hand, two long ends in your right hand, and one end of the medium rope in each hand.

Be sure that the middle portion of the short rope is being held in your left hand, as shown in Figure 9.55. If you need to, shift the ropes up so the middle is residing in your left hand.

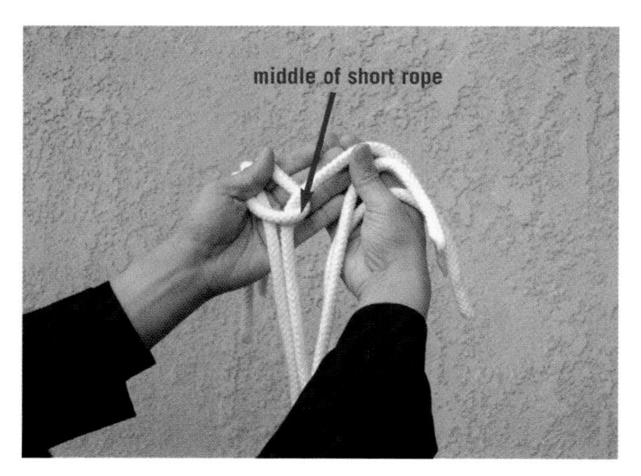

Figure 9.55
If you need to, pull up on the ropes so the middle of the short rope is shifted into the left hand.

From the audience's perspective, it will look as if you are holding the short, medium, and long ropes between your two hands, as shown in Figure 9.56.

Figure 9.56
This is what it looks like to the audience.

After you're certain that you have the correct grip, slowly pull your hands apart. It will appear that the short rope is stretching while the long rope is contracting. The illusion is very convincing. When you're finished pulling, it will appear that you have three ropes of equal length, as shown in Figure 9.57.

Figure 9.57
The ropes appear to have changed into three ropes of equal size.

Figure 9.58 offers a view from the back to show you the current configuration of the ropes.

Figure 9.58
A view from the back.

As a convincer, you can tug on the ends to show that the ropes are solid.

To bring the ropes back to their original lengths, maintain your grip on the ropes with your left hand and loosely wrap all three ropes around your left hand. Simply find the individual lengths of ropes and remove them, one by one, to show that each rope has returned to its original length. This is shown in Figure 9.59.

Figure 9.59
Loosely wrap the ropes around your hand and remove the ropes one by one.

Next Steps

After our chapter on stand-up magic, you can move onto Chapter 10, which discusses street magic. Here you'll learn edgy close-up tricks that work in unpredictable environments.

Magic for the Street

STREET MAGIC IS AMONG MAGIC'S OLDEST TRADITIONS. Since the Middle Ages, and even before, magicians have plied their trade on the streets. Sometimes, it was to gather and entertain crowds and request remuneration in return. Other times, it was to provide a diversion for other "artists" of the period, which included pickpockets, who could use the opportunity to ply their particular trades.

In the last decade, the definition of street magic has shifted, mostly because of magician David Blaine. Back in the nineties, Blaine appeared from seemingly out of nowhere on a television special called "David Blaine: Street Magic." What Blaine presented was not traditional street magic, but seemingly impromptu close-up magic that he performed for spectators, all with the cameras rolling. In the end, the best segments, which featured excited participants, were shown on his special.

David Blaine began with street magic and now performs stunts on national television.

Indeed, Blaine created a sensation and more specials soon followed. Most memorable in that first show was a levitation where he raised himself almost two feet off of the ground.

With an eye towards tradition and close-up magic that you can perform on the street by walking up to spectators and "showing them something," we present street-style tricks: close-up effects as well as classics that have long been performed by street entertainers. And by the way, if you want to learn to levitate, you can in Chapter 12.

Mean Streets

OF COURSE, PERFORMING on the streets doesn't literally mean that you will be performing in an urban environment. You can end up performing in parks, sidewalks, malls, and more. For the most part, your props have to be minimal and easy to carry around and preferably on your body without the need to carry a bag. Because the street environs are so unpredictable, you'll need tricks that work when you are surrounded by eager spectators.

The Mysterious Kiss

Here's a card trick that can play well in almost any situation. It's one that also encourages lots of audience participation.

©istockphoto.com/David Kneafsey

Effect

A card is selected and returned to the deck. You ask a spectator to blow a kiss at the deck. You look through the deck and tell the spectators that you think you have found her card. You name the card and the spectator verifies that it is indeed her card. When you turn the card around, you show that it has lip prints on it. You can give the card away as a souvenir.

Secret

You force a card and have a duplicate with a set of lips already printed on it.

Materials

A deck of cards.

An extra card, preferably a "two" or "four" that has lots of white space.

Stickers that depict a lip print, as shown in Figure 10.1, or an ink stamp that imprints an image of a kiss.

Figure 10.1
You'll need stickers or a stamp that can imprint a pair of lips.

Skills

The ability to force a card. For our purposes here, we'll use the cross-cut force.

The ability to false shuffle or cut the deck (optional).

Preparation

Take your extra card and fix a "kissing lips" sticker onto the white part of the card's face, or use your stamp to accomplish the same thing. Figure 10.2 shows a gimmicked card. (We'll use the two of hearts in this example.)

Figure 10.2
A gimmicked "kiss" card.

Place the "kiss" card on the top of the deck and its ungimmicked duplicate on the top of everything, as shown in Figure 10.3.

Figure 10.3
Place the gimmicked "kiss" card on top and the matching normal card on top of everything.

Performing the Trick

Casually show the faces of the cards to spectators and be careful not to show the top cards.

Perform a false shuffle or cut if you like.

Perform the cross-cut force as explained in Chapter 4. After the spectator has cut the deck, you can talk about the power of a kiss or whatever topic you dream up.

Ask the spectator to take the card that they "cut" to, which will be the top card on the bottom half. Ask the spectator to show it to others and remember it. Spread the deck and allow the spectator to place the card back into the deck at any location.

Genuinely mix the deck so the spectator's card is lost.

Ask the spectator to blow a kiss at the deck. Depending on your personality, you can play this one up for laughs.

After talking about the power of a kiss, turn the cards so they are facing you and look through the deck for your force card with the lips. When you find the two of hearts with the lip prints, bring out the card but don't yet reveal it to spectators, as shown in Figure 10.4.

Figure 10.4
Bring out the card with the lip print. For now, hold it so you don't yet reveal this to spectators.

Ask the spectator if her card was the "two of hearts." When she verifies this, turn her card around to show that you have the two of hearts and there are lip prints on the card.

Card to Wallet

This is a variation on the "card to impossible location" plot that is similar to a trick that was explained in Chapter 6. Unlike that effect, this one will use an ordinary deck and not a Svengali deck and allow you to produce a selected card from your wallet.

Effect

A spectator's selected card vanishes from the deck and is found inside of your wallet.

Secret

You force a card and its duplicate is already in your wallet.

Materials and Requirements

A deck of cards.

A second card that matches the card that you plan to force.

A wallet that has a free pocket to hold the playing card.

Skills

The ability to force a card (any method).

False shuffling or cutting a deck (optional).

Preparation

Place the extra matching card in your wallet, as shown in Figure 10.5. Place the wallet into your pocket.

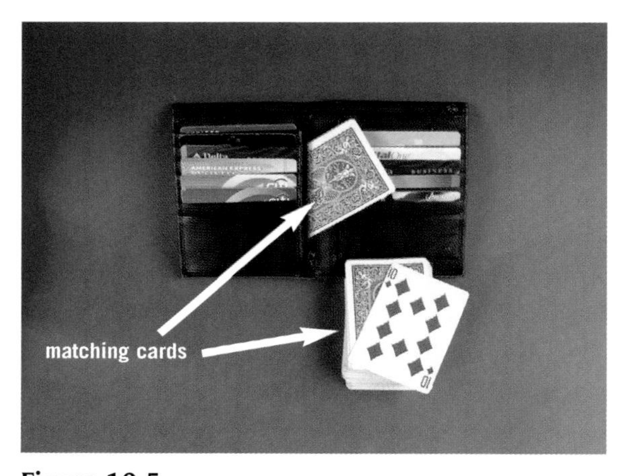

Figure 10.5
The second matching card goes into a pocket in your wallet.

Position your force card in its designated location for your force.

Performing the Trick

Bring out the deck and casually show the faces of the cards to spectators.

Perform a false shuffle or cut if you like.

Force your card. If you have a table, you can use the cross-cut force as taught in Chapter 4. If you don't have a table to work on, you can easily use the Hindu force, also explained in Chapter 4.

Ask the spectator to show his selected card to others and remember it. Spread the deck and allow the spectator to place the card back into the deck at any location. Genuinely mix the deck so the spectator's card is lost.

Tell the spectators that the trick is now complete. Put the cards into their box and put away the deck.

Show your hand clearly empty and then reach into your pocket and pull out your wallet.

Slowly open your wallet and bring out the card so it's face down.

Ask the spectator to name his card.

Turn over the card to reveal that the selected card has amazingly traveled to your wallet, as shown in Figure 10.6.

Figure 10.6
The spectator's card has flown to your wallet.

The Rising Ring

This is a levitation effect that you can perform just about anywhere. Here, a finger ring, which can be borrowed, seems to defy gravity by climbing up a rubber band on its own accord.

Effect

A ring is placed onto a rubber band and appears to slowly rise, as if defying the laws of gravity, as shown in Figure 10.7.

Figure 10.7
Defying gravity with a ring.

Secret

You hold the rubber band under tension and slowly release the band. As the band releases, it carries the finger ring along with it.

Materials

A rubber band. This one is better with a larger rubber band.

A finger ring.

Preparation

None.

Performing the Trick

Hook the rubber band on the first finger of your left hand and thread the finger ring onto the rubber band, as shown in Figure 10.8.

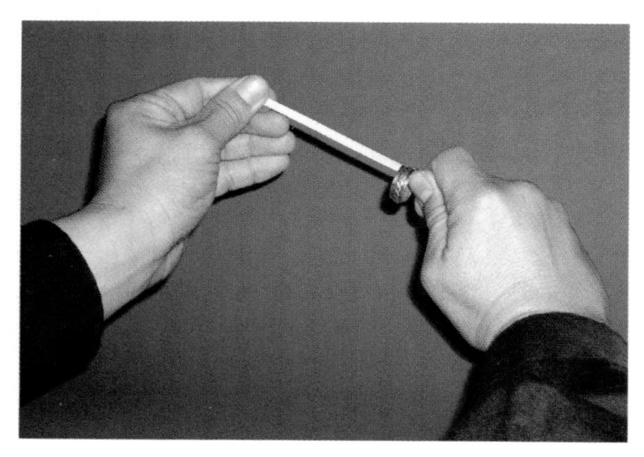

Figure 10.8
Hold the rubber band with your left first finger and thread the finger ring onto the rubber band.

Using your right hand, grab a good portion of the rubber band so most of it is held in your hand. Depending on the length of the rubber band, you'll want almost half of the band to be in your hand. Gather this slack in your hand as you talk with spectators, as shown in Figure 10.9.

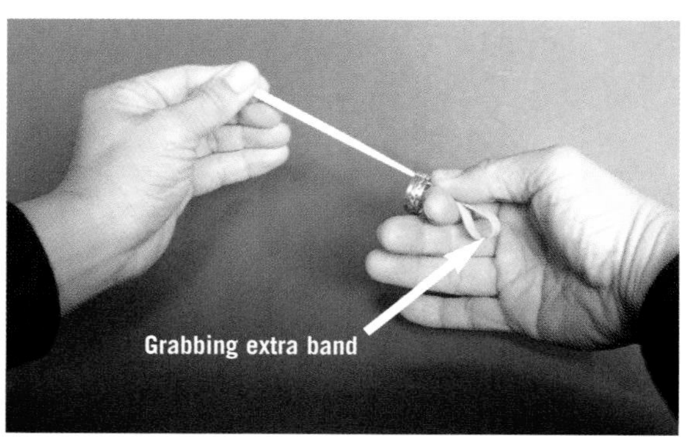

Figure 10.9
Grab a good portion of the rubber band in your right hand.

Pinch the middle of the rubber band using your right first finger and thumb and stretch the band. Bring your right hand down so the rubber band is at an angle and allow the finger ring to fall so it's resting on your right hand, as shown in Figure 10.10.

It's important that you practice setting up this trick so it's smooth. If you take too much time gripping the extra rubber band or look as if the process is taking too much effort, spectators will suspect something.

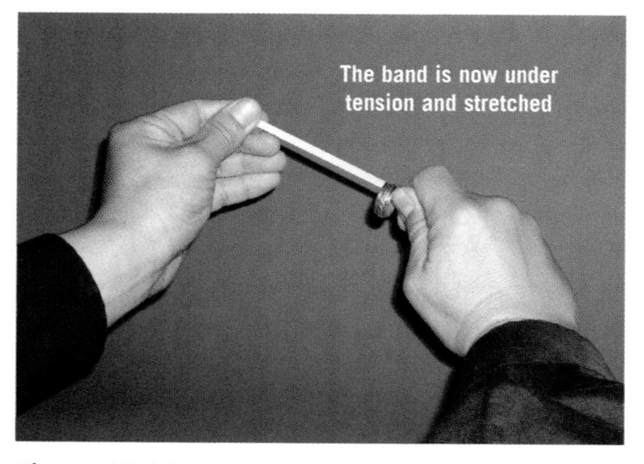

Figure 10.10
The starting position of the ring rise.

Slowly release the grip with your right first finger and thumb so the band is slowly released. As you release the band, the ring travels with it. To spectators, it will look as if the finger ring is slowly rising up the band, while it's actually moving with it. This is shown in Figure 10.11.

Figure 10.11

By slowly releasing the rubber band with your right finger and thumb, the ring will appear to travel up the band.

It's important that you maintain the same distance between your hands. You don't want to move your hands, but simply allow the rubber band to move the ring. If you're not getting much ring movement, you need to establish more tension in the beginning by pulling more of the band into your right hand. You'll need to experiment with this.

When the ring can't move anymore, simply bring your hands together, release the band with your right hand and remove the ring.

This trick works well as a closer to the rubber band tricks explained in Chapter 7.

Card on Wall

Here's a classic of magic that will allow you to find your spectator's card in a spectacular fashion, by throwing the deck against a wall and causing the selected card to stick to it.

Effect

A spectator selects a card and returns it to the deck. You toss the deck against a wall and as the cards fall onto the ground, everyone discovers that the spectator's card has stuck to the wall.

Secret

You use a loop of clear adhesive tape to secretly adhere the spectator's card to the wall.

Skills

The ability to bring a selected card to the top of the deck. While you can use any method you like, we recommend either the double-cut or Hindu control. Both of these are taught in Chapter 4. While it's not as clean, you can also use a keycard control.

Materials

A deck of cards. Since your cards will end up on the ground and get grimy, it's best to use an older deck or dedicate a deck to this effect.

Clear adhesive tape.

Preparation

Take about two inches of clear adhesive tape and make a loop with the sticky side on the outside. Place the tip of the little finger of your left hand into the loop. You'll have to hold and hide this loop until you're ready to throw the cards onto the wall, as shown in Figure 10.12.

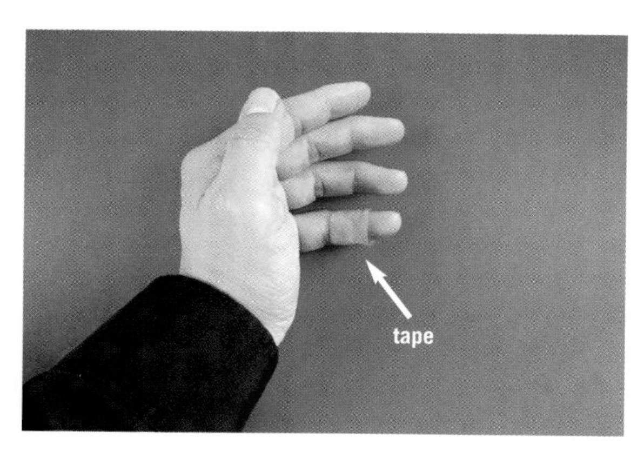

Figure 10.12
Place the loop of adhesive tape around your left little finger.

You'll have to ensure that you have the right type of wall to perform this trick. A brick or stucco wall may be too rough to allow the tape to stick to it. For this reason, it's best to test your deck with the wall that you plan to use and make sure that the card will stick.

> **Hint**
>
> To test a wall beforehand, wrap a rubber band around your deck of cards but leave one card, the one that will stick to the wall, out of the rubber band. This way, you can throw the deck against the wall and not have to pick up the other cards.

Performing the Trick

Take out the deck of cards and allow a spectator to select a card.

Using your favorite method, allow the spectator to return the card to the deck and then bring the card to the top. Here you can use a key card, or a double-cut, as explained in Chapter 4.

Turn over the deck so it's face up in your hands, and as you hold the deck with both hands, with your left little finger, press the loop of tape on the back of the top card, the selected card, as shown in Figure 10.13. Get the tape as close to the center of the card as you can.

Grab the deck with your right hand to prepare to throw the deck. If you are throwing past the spectators who might be able to see the top card, cover the back of the deck with your left hand so the spectators can't see the tape. This is shown in Figure 10.14.

Figure 10.13
Press the loop of tape onto the back of the top card. Note that this is the view from underneath the deck, and the right hand, which would be helping to hold the deck, has been removed.

Figure 10.14
Grab the deck with your right hand to prepare for the throw.

Ask the spectator to name his card and then toss the cards with an overhand throw towards the wall. Make sure that the top card is the first to hit the wall and that the other cards land on top of it to help it stick. It's the momentum of the rest of the pack that causes the selected card to stick to the wall. The other cards will fall down and the spectator's selected card will remain stuck to the wall.

Should the card not stick to the wall, you can try and salvage the effect by picking up the cards and setting up the "throw" again with the "taped card" on the back. This time, instead of throwing the cards against the wall, simply press the entire deck to the wall and make sure that the back card sticks, and then remove the deck to show the spectator's card stuck to the wall.

The Great Escape

Escapes are a traditional part of a street performer's repertoire. And some performers specialize in escapes from strait jackets and other bindings. Here's an escape that uses low-cost items that you can find at your local pet store.

©istockphoto.com/Kris Hanke

Effect

Your wrists are bound with a chain and a lock, as shown in Figure 10.15.

You turn your body and you've quickly released yourself from the chains. The spectators can examine the chains and locks and will find them to be legitimate. And the lock is still secure and unopened.

Figure 10.15
Your wrists are tightly bound with a chain and lock, yet you escape.

Secret
It's all in how you tie yourself up in the chain.

Materials
A small lock with a key.

A dog collar chain that you purchase from your local pet store, the kind with rings at both ends. Depending on the size of your wrists, an 18" chain will probably work for most. It's best to go through the instructions here and thoroughly understand how the chain is bound to your wrists, even practicing with a rope.

Later, you can go to the pet store and try the different chains to see which one works best for you. Yes, you'll get some strange looks from other customers. But hey, we all have to suffer for our art.

By the way, I purchased the dog collar chain shown in the pictures here at a local dollar store as well as the lock. The equipment for this trick was a little over two dollars.

Preparation
None.

Performing the Trick
Open the lock and hand it to a spectator to hold until you need it. You can place the key in your pocket.

Start by creating a loop with one of the rings and the length of chain and place your left hand through it. Your hand is turned with its thumb side up. As you tighten the loop, be sure that the ring is resting near the top of your wrist, as shown in Figure 10.16. This is the key to the trick.

Figure 10.16
Make a loop with the ring and chain and place your left hand through it. Note how the ring is resting near the top, upper part of the wrist.

Allow the chain to hang down from the ring, as shown in Figure 10.17. This is important as it provides the necessary slack so you can make your getaway. Notice how the ring remains near the top of the left wrist.

Figure 10.17
Allow the chain to hang down.

Place your right wrist against your left wrist. Like your left hand, your right hand is turned with its thumb on top, as shown in Figure 10.18.

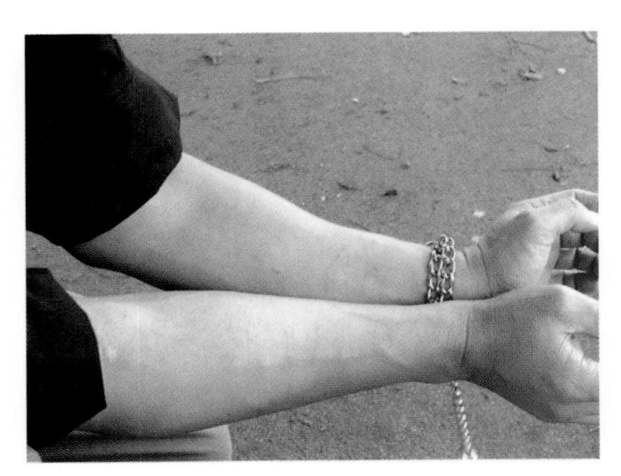

Figure 10.18
Rest your right wrist against your left wrist.

Invite a spectator to wrap the remaining length of chain around your right wrist and then lock the end of the chain to the ring.

Be sure to hold your hands in place and push against the chains. While it appears that you are securely shackled, you can easily get out and you don't want to accidentally release yourself. You will be shackled as shown in Figure 10.19.

Figure 10.19
Your wrists are shackled.

To escape, turn your body so you're facing away from spectators to hide the secret. By bringing your right wrist up and over the left wrist, you'll find that there's enough slack for you to quickly slip your right wrist out, as shown in Figure 10.20.

Figure 10.20
Bring your right wrist up and over your left wrist to create some slack to slip out your right wrist.

Pull your left hand free by expanding its loop. Pull both hands out and turn around and show spectators that you have escaped, as shown in Figure 10.21.

Figure 10.21
Despite a legitimate chain and lock, you've quickly escaped.

If you like, you can make your escape under the cover of a cloth or jacket. Just have a spectator drape your hands so you can make your escape. You won't have to turn if you choose to escape while under cover.

Card Monte

The "Three Card Monte" is a well known street swindle that uses three cards. It takes lots of skill and experience to throw the real Monte, but we'll teach a simple effect that is similar in spirit.

Effect

Three playing cards are shown to spectators: two black number cards and a red queen. You ask spectators to follow the queen. The three cards are laid down on the table and when a spectator points to the card that he thinks is the queen, he discovers that the queen has turned into a black number card. You reach into your pocket and produce the queen.

Secret

You create a gimmick that makes a black number card appear to be a queen.

Materials

Two like red queens, either two queen of hearts or two queen of diamonds.

Three black number cards. For this, it's best to use high numbers such as eights, nines, or tens.

Adhesive tape.

Scissors or a paper cutter.

Preparation

Cut one of the red queens in half and tape the two halves of the queen together, as shown in Figure 10.23.

Figure 10.23
Cut one of the red queens in half and tape the two halves together.

You now have a gimmick that will fit over a black number card and make it appear to be the red queen, as shown in Figure 10.24. Fold the gimmick in half and slip the black number card in between the two halves of the queen.

Figure 10.24
The black number card can now look like a red queen.

Place the second, matching red queen into your pocket.

Stack the cards as follows: A black number card, the black number card with the queen gimmick on top, and the third black number card.

Performing the Trick

Bring out your stack of cards and pull out the top two cards and display them in your left hand, as shown in Figure 10.25. Of course, the middle card that appears to be a red queen is actually a black number card with your gimmick on top.

Figure 10.25
Display the cards.

Push the cards together and turn them face down, as shown in Figure 10.26.

Figure 10.26
Push the cards together and turn them face down.

Place the stack of cards into your right hand, as shown in Figure 10.27. Notice how this turns the cards around. The gimmick is now closest to your right wrist.

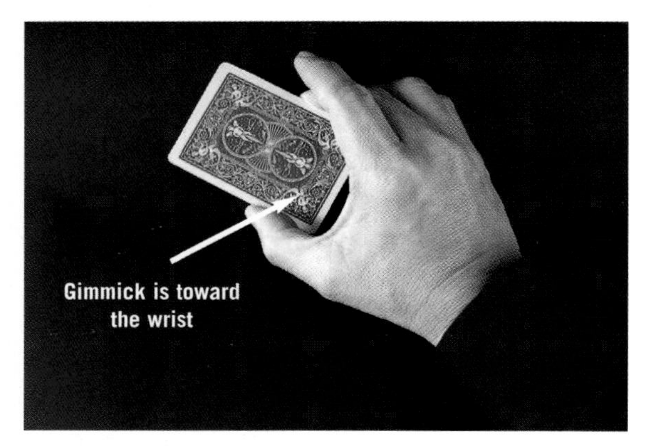

Figure 10.27
Place the stack into your right hand.

Without showing the cards, pull out the bottom card and rest it on the table. Pull out the second card, which is the black number card, and place it onto the table next to the first card. Keep the queen gimmick hidden in your right hand and lay the top card, another black number card, onto the table, as shown in Figure 10.28.

Figure 10.28
Keep the queen gimmick hidden in your right hand as you pull out the last card.

Ask a spectator which card he thinks is the queen. Have him turn over the cards to show that the queen has turned into a black number card, as shown in Figure 10.29.

Figure 10.29
The queen has turned into a black number card.

While spectators are turning over the cards, reach into your pocket and leave the queen gimmick there and pull out the real queen. The queen has mysteriously traveled to your pocket.

The Bottle Caps

Here's an effect that's seemingly made for the street. It's fast and visual and can be performed on a table, bench, or even a sidewalk. And it feels like an edgy street hustle.

Effect

Four bottle caps are laid out on the ground. The caps disappear and reappear from under your hands until they're magically grouped together. This routine is based on a classic one that magicians call the "Chinese Assembly."

Secret

You have five bottle caps that allow you to position one cap while secretly moving another. Spectators only see four of the bottle caps.

Materials

Five bottle caps.

Preparation

You can do this at any time once you master the palm.

You'll need to learn how to pick up a bottle cap using the palm of your hand. The cap goes into the center of the palm of your hand, as shown in Figure 10.30.

Figure 10.30
Positioning the bottle cap.

By squeezing your thumb in a bit, you can hold the cap and raise your hand off of the table. This is shown in Figure 10.31.

Figure 10.31
With practice, you can hold a bottle cap in your hand and lift it off of the table. The inset shows the hand palm down while holding the bottle cap.

Magicians call this palm a "classic palm," which is most commonly performed with a coin. Because of the higher edges and scalloped ridges of a bottle cap, a bottle cap is easier to palm than a coin. The practice that you put in now will help you later when you want to learn how to classic palm a coin.

> **Hint**
>
> If you're having trouble picking up the metal bottle caps, try using plastic caps from water bottles that are higher. You may have an easier time handling these.

Performing the Trick

Bring out four bottle caps and palm one in your right hand. Lay out four bottle caps as shown in Figure 10.32. Note that we've labeled the positions A, B, C, and D.

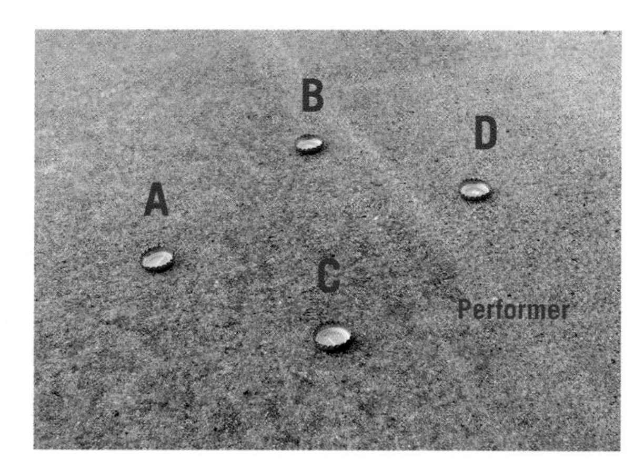

Figure 10.32
The starting position for this trick.

Place your left hand on C and your right hand on D. Wiggle your fingers and palm bottle cap C in your left hand and drop the bottle cap that's palmed in your hand at location D, as shown in Figure 10.33. Be careful that you don't allow the bottle caps to hit and make a noise.

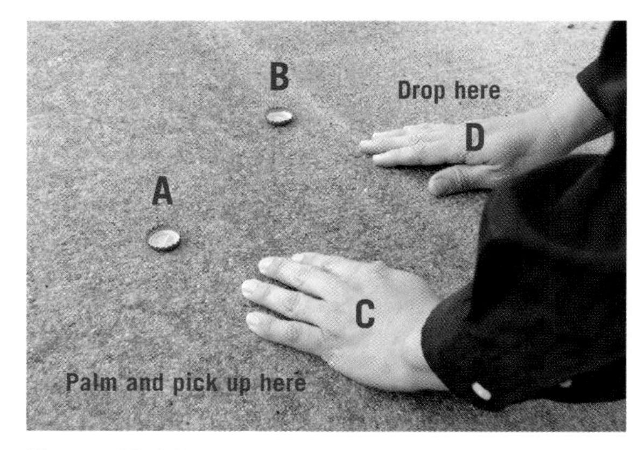

Figure 10.33
Pick up bottle cap C and drop your palmed bottle cap at D.

Remove your hands. To spectators, it will look as if the bottle cap at C has vanished and reappeared at location D with the bottle cap already there. The bottle caps will look as they do in Figure 10.34.

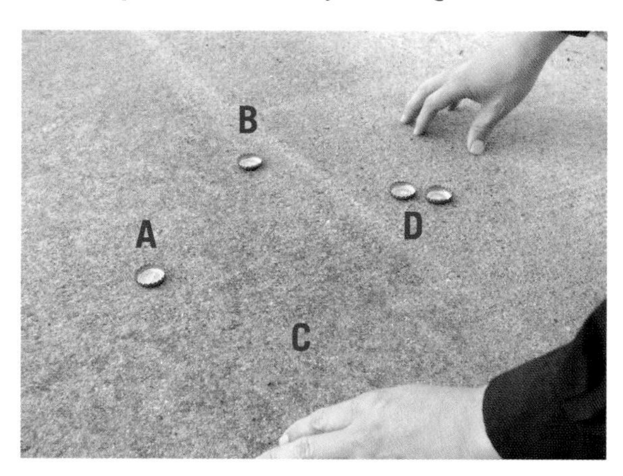

Figure 10.34
The first stage.

> **Reality Check**
>
> Unknown to spectators, you now have a bottle cap palmed in your left hand.

Cross your arms and place your left hand on D with its palmed cap and right hand on A, as shown in Figure 10.35.

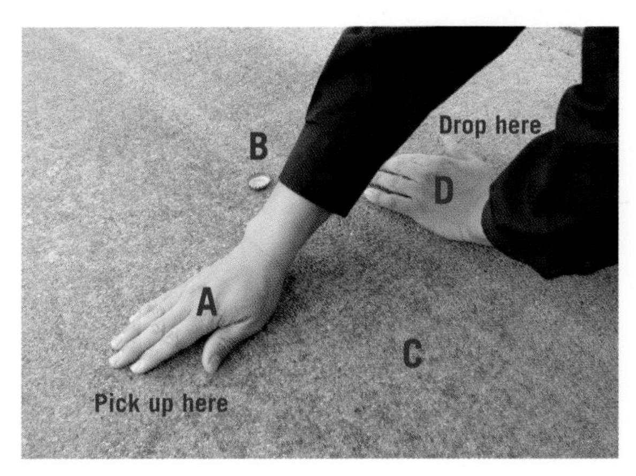

Figure 10.35
Place your left hand on D and right hand on A.

Wiggle your fingers and drop the bottle cap that's in your left hand at D and pick up and palm the bottle cap at A in your right hand. Remove your hands and you'll have the situation shown in Figure 10.36.

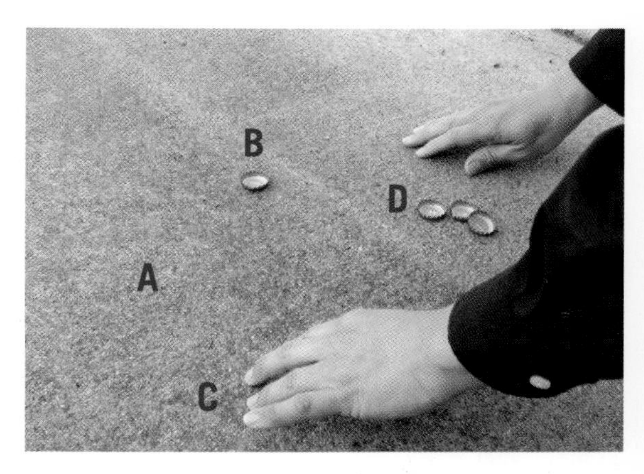

Figure 10.36
Stage two.

> **Reality Check**
>
> Unknown to spectators, you now have a bottle cap palmed in your right hand.

Place your left hand on B and right hand on D, as shown in Figure 10.37.

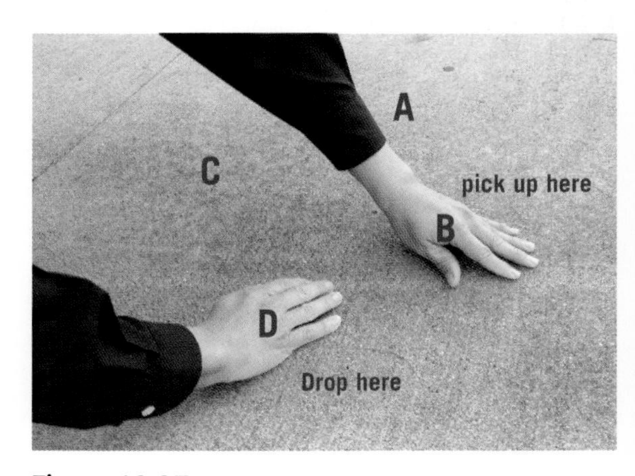

Figure 10.37
Place your left hand on B and right hand on D. Note that we've switched angles for clarity.

Wiggle your fingers. With your left hand, pick up the bottle cap at B, and with your right hand, drop your palmed bottle cap at D. When you remove your hands, four bottle caps will be grouped at D, as shown in Figure 10.38.

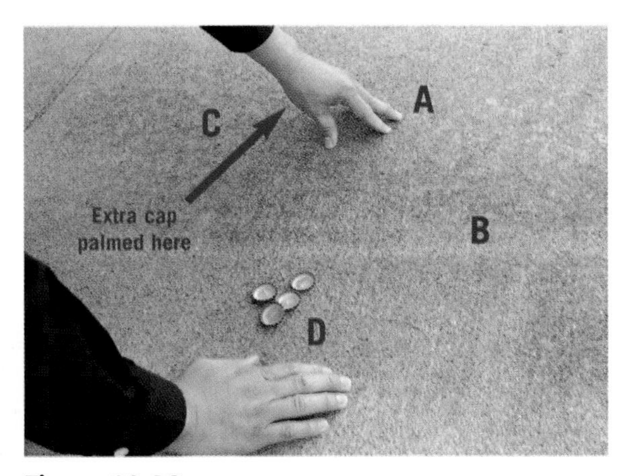

Figure 10.38
At the end, four bottle caps will be grouped at D.

Keep the extra bottle cap palmed in your hand and try to relax your hands so they don't draw attention. When your audience is relaxed, you can place your hands in your pockets and get rid of the extra bottle cap. Or you can gather all of the bottle caps and dump all of them, including the palmed one, into your pocket.

Cups and Balls

If this trick isn't the oldest in magic, it's definitely one that's been around for a long time. It was once thought that the cups and balls were depicted in murals in Egypt's pyramids, but based on what we now know, this is probably not the case. However, despite the lack of evidence, most experts feel that the trick was probably around during the age of Pharaohs.

To perform cups and balls, you'll need a set of inexpensive cups and balls. What makes these cups different from standard cups is a little, built-in shelf that allows the cups to stack, but can accommodate a hidden ball, as shown in Figure 10.39.

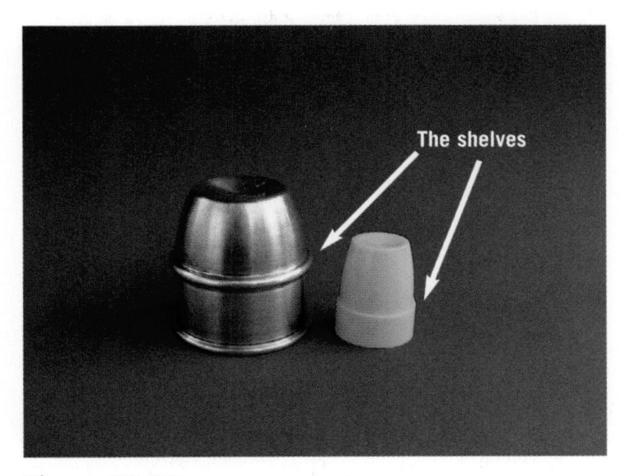

Figure 10.39
The cups have a shelf that allows a cup to stack and hide a ball inside.

Serious magicians often spend hundreds, or even a thousand dollars or more on their cups. However, to learn the basics of the cups and balls, you can use an inexpensive plastic set that will only set you back about five dollars from a magic dealer.

Depending on the surface that you're working on, you may need balls that are soft and quiet. Magicians often used embroidered balls, but you can also use pom poms or fashion balls out of foam. If you purchase a set of cups and balls, it will usually come with an appropriate set of four balls that you can use.

The routine that we offer here is a basic one that will start you on your way with the cups. If you become serious about the effect, there are entire books that are written about it. The expert tricks require lots of sleight-of-hand and a magician's cups and balls routine is often the mark of one's skill in sleight-of-hand.

Effect

Three cups and three balls are shown to spectators. Two balls magically pass through the solid cups, one at a time, and a third ball magically moves from between two cups and under a third cup to join the other two.

Secret

You have an extra ball that allows you to perform the penetrations and transposition.

Materials

Three cups for cups and balls.

Four balls.

Preparation

For explanation purposes, we're going to designate the cups as cup A, B, and C. To start, you'll stack the cups together with the mouths up. Beforehand, place one ball into the mouth-up A. Then stack B and C on top of A. Place the remaining three balls into the mouth of C. You're ready to go.

> **Hint**
>
> Place stickers or tape notes to your cups so it's easier to follow along with the instructions.

Performing the Trick

Bring out the cups and balls and pour the balls from cup C onto the table, as shown in Figure 10.40.

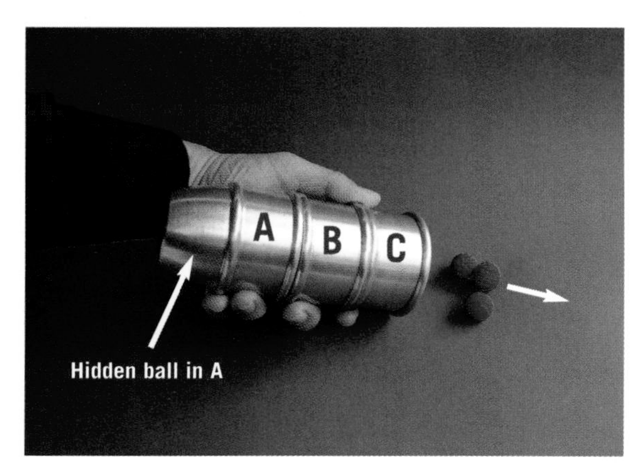

Figure 10.40
Pour the three balls from cup C onto the table.

Take cup A from the bottom of the stack, and quickly move it over and rest it, mouth down on the table. During this process, you'll quickly flip cup A, which contains the extra ball, until it's resting mouth down on the table. There's a definite knack to this. Just use momentum to keep the ball in the cup until the cup is firmly resting in place. This is shown in Figure 10.41.

Repeat the process with the remaining two cups, turning them over and resting them mouth down on the table. With all of the cups mouth down on the table, rest a ball in front of each cup, as shown in Figure 10.42.

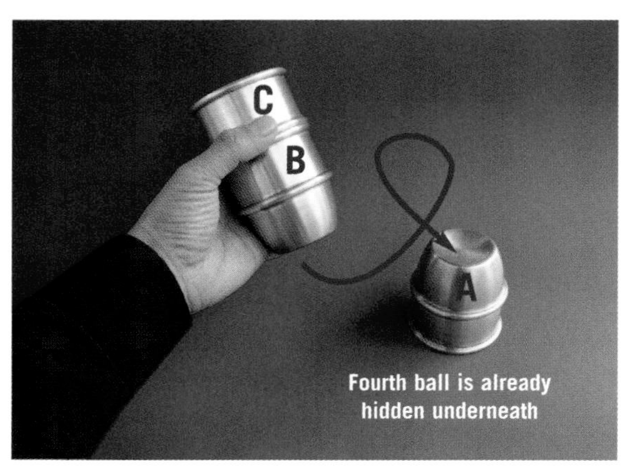

Figure 10.41
Remove cup A from the bottom of the stack and place it mouth down on the table.

Figure 10.42
Starting position for the trick.

> **Reality Check**
>
> The rightmost cup, cup A, already has the fourth ball resting under it.

Place a ball on top of cup B and rest cup C on top, as shown in Figure 10.43.

Figure 10.43
Rest a ball on top of cup B and rest cup C on top.

Place the stack of cups B and C on top of cup A, as shown in Figure 10.44.

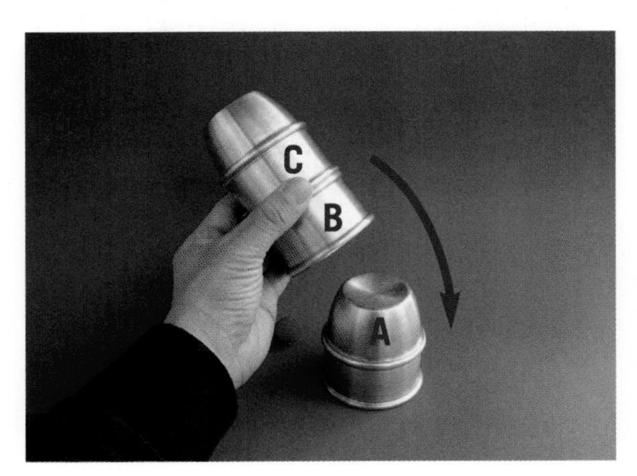

Figure 10.44
Place the stack of B and C on top of cup A.

Lift the entire stack to reveal a ball under cup A, as shown in Figure 10.45.

Figure 10.45
Lift the entire stack to reveal a ball.

Reality Check

The rightmost cup, cup C, now has two balls resting under it.

Turn the entire stack mouth up. Grab the bottom-most cup, cup C, and rest it mouth down on top of the single ball, as shown in Figure 10.46.

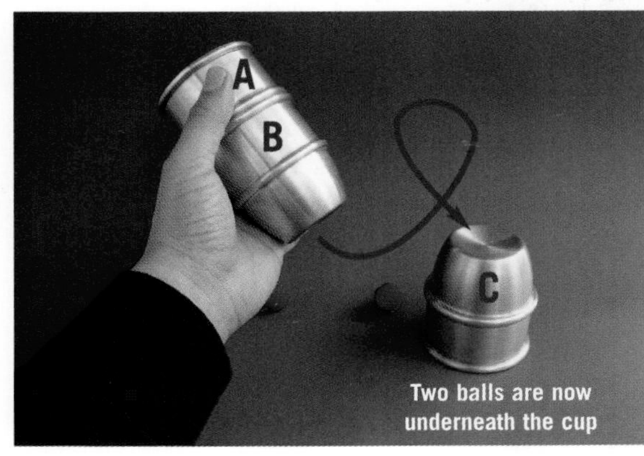

Figure 10.46
Place cup C mouth down on top of the single ball.

Pick up a ball and rest it on the top of cup B and place cup A on top, as shown in Figure 10.47.

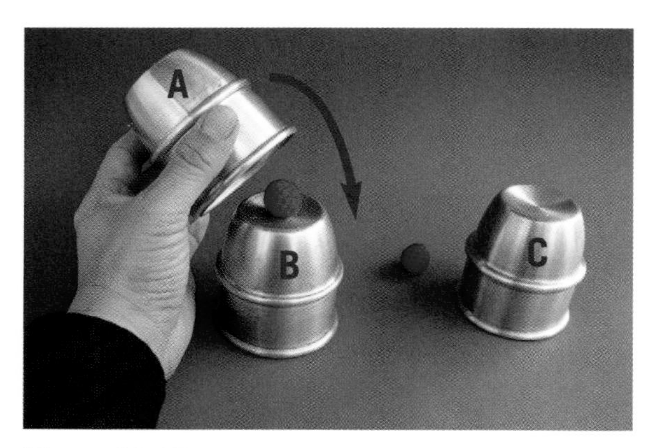

Figure 10.47
Ready to perform the second phase.

Place the stack of cups A and B on top of cup C, as shown in Figure 10.48.

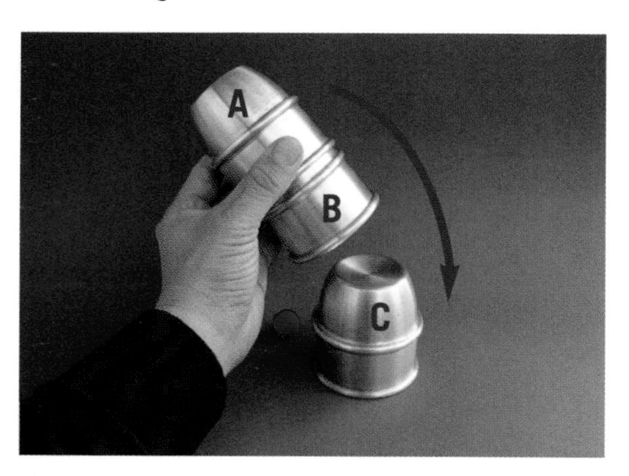

Figure 10.48
Place the stack of cups A and B on top of cup C.

Snap your fingers and lift the entire stack of cups to reveal that the second ball has joined the first, as shown in Figure 10.49.

Figure 10.49
Reveal that a second ball has joined the first.

Turn the stack of cups mouth up and grab the bottom-most cup, A, and rest it mouth down on top of the two balls.

> **Reality Check**
>
> Cup A now has three balls resting under it.

Rest cup B mouth down on the table and place a ball on top of it. Rest cup C on top of the ball and cup B, as shown in Figure 10.50.

Figure 10.50
Ready for the third phase.

Snap your fingers and touch the two cups to the left and then the rightmost cup, and then, lift the rightmost cup to reveal three balls underneath, as shown in Figure 10.51.

Figure 10.51
The third ball has mysteriously joined the other two.

Stack the cups by taking the two cup stack and turning it mouth up. Take the rightmost cup and place it on the top of the stack, and then place the three balls into the mouth of the topmost cup. You're ready for another performance. The extra ball is already in place inside of the bottom cup.

Next Steps

After our chapter on street-style magic, in the next chapter, you'll learn magic that you can perform at parties and other social gatherings.

©istockphoto.com/Renee Lee

11

Party Magic

FOR BEGINNING MAGICIANS, parties are excellent places to perform magic. There's a gathering of people socializing and out to enjoy themselves, and many times, someone who knows that you're into magic may just ask you, "how about a trick?" For these situations, we offer this chapter.

Conditions for Party Magic

SO WHAT CONDITIONS CAN YOU expect at a party? You'll need to be flexible, as you can't count on any specific performance situation. The tricks that you perform have to be those that can work from a variety of angles. In many party situations, it's hard to situate yourself in a position where there's no one standing next to you or behind you.

While close-up magic is great for parties, you'll want effects that can be easily seen as you may be performing for five or more people, perhaps for even a room full of spectators.

You'll want to perform tricks that feature lots of interaction. The more people that you can involve in an effect, the better. And since you may not have your specific magic props with you, you'll want to be able to perform magic on the spot with whatever happens to be around.

Visualize the Tricks

Before you begin to learn the various tricks, you may find it helpful to view them in action on the accompanying DVD.

For this, please refer to the "Party Magic" video that demonstrates each trick. After watching the video, you may discover that there are tricks that you want to learn first and move directly to the appropriate section.

Dark Magic

Magicians have long used secret codes to communicate with their assistants. Perhaps the most famous of these routines is known as "Second Sight," which was performed by the famous French magician, Robert-Houdin, with his son Emile, back in the 19th century.

In this trick, "Dark Magic," you communicate with a friend through a simple secret code, which reveals an object that was chosen when you are out of the room. This one is baffling and great for parties.

Effect

You step away from the room and a group of people select an object. You walk back in and without a word, your assistant simply points to the objects, one by one. When the assistant points to the selected object, you correctly identify it as the one that was chosen by the group. There appears to have been no communication between you and your assistant.

Materials and Requirements

A room with a variety of objects. Alternatively, you can use a table that has a bunch of objects on it, just as long as the objects meet the criteria of the secret code, as shown in Figure 11.1.

Figure 11.1
All you need is a variety of objects.

Figure 11.2
Your assistant points to a black object or the color black on an object.

Secret

When your assistant points to a black object or the color black, the next object that he touches will be the selected object.

Preparation

This one can be done at any time and there is no preparation.

Performing the Trick

Tell a group of people to agree on a selected object in the room. Leave the room so they can make their decision. When they have agreed on an object, someone brings you back.

Your assistant begins to silently point to different objects, one at a time.

When your assistant points to a black object or the color black on an object, you know that the next object will be the chosen one, as shown in Figure 11.2.

When your assistant points to the object after the black object, you tell the crowd that this is the object that they chose, as shown in Figure 11.3.

Figure 11.3
The next object is the chosen object.

It's important that your assistant do everything possible to point at each object in the exact same way, without variance, because everyone will be looking for some physical signal.

While it's usually not a good idea to repeat a trick, this one can be repeated to good effect. If you choose to perform it a second or even third time, ask your assistant to mix up the order of the revealed objects because some spectators will suspect that you and your assistant are counting to a certain agreed number.

The Red Card

This intriguing card trick offers a play on words. You'll find this one easy to perform and the revelation is not only baffling, it has a fun surprise.

Effect

You bring out a blue-backed deck and have a spectator select a card. You tell him that he has selected a "red card." The spectator, who has selected a club or spade, a "black card," tells you that you are incorrect. But you ask the spectator to turn his card over to show that it's the only red-backed card in the blue deck.

Materials and Requirements

A blue-backed deck of cards.

One red-backed playing card from another deck that is a spade or club.

You'll need to perform this one on a table.

Secret

You force the only red card using the cross-cut force.

Skills

The cross-cut force (Chapter 4).

Preparation

Place the red-backed card face down on the bottom of the blue-backed deck, as shown in Figure 11.4.

Figure 11.4
Secretly place the red-backed card face down on the bottom of the blue-backed deck.

If you like, you can remove the blue-backed card of the same value as your red card, as some spectator may want to check for this. However, this isn't important and doesn't detract from the trick.

Performing the Trick

Bring out the blue-backed deck and perform a false cut. Be careful that you don't expose the red-backed card on the bottom of the deck.

Rest the deck on the table.

Ask the spectator to cut the deck. The spectator cuts half of the deck off the top and places it onto the table. Take the original bottom half and rotate it ninety degrees and place it on top of the other half, in classic "cross-cut" fashion, as shown in Figure 11.5.

Figure 11.5
Execute the cross-cut force.

Talk briefly about the various suits in a deck: clubs, hearts, spades, and diamonds and how the clubs and spades are known "black cards" and the hearts and diamonds are known as "red cards." You now tell the spectator that you will read her mind.

Ask the spectator to lift the top half of the deck and look at the bottom card to which she cut.

> **Reality Check**
>
> Because you have performed the cross-cut force, the spectator will be looking at the original bottom card, the red card that you set up earlier.

Have the spectator close the deck.

Pretend to read the spectator's mind and then state that she chose a "red card."

When the spectator tells you that she did not choose a red card, you insist that it's a red card.

At this point, turn over the entire deck so that it's face up and ask the spectator to tell you the name of her selected card. Find the card and set it aside. Be careful not to reveal that it's a red-backed card, as shown in Figure 11.6.

Figure 11.6
Find the spectator's card and set it aside.

Show that the rest of the deck consists of blue-backed cards, as in Figure 11.7.

Figure 11.7
Show the deck to consist of blue-backed cards.

Then turn over the selected card to show that it's the only red-backed card, as shown in Figure 11.8.

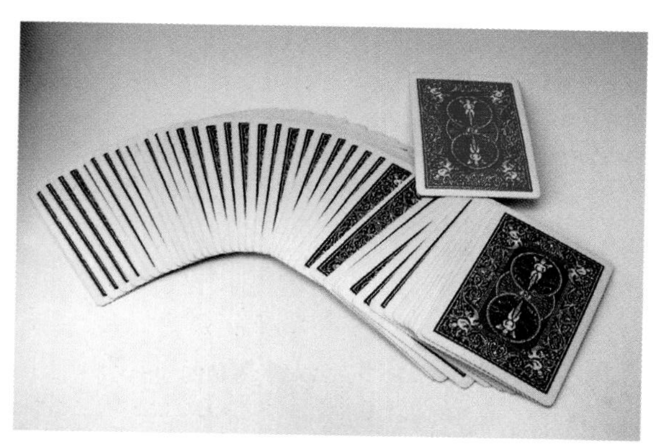

Figure 11.8
The spectator has chosen the only red-backed card in a blue-backed deck.

Touch a Card

This is another effect where a freely chosen object is communicated to you by an assistant. In this case, it's a card that was selected from several that are resting on the table. As with "Dark Magic," which was taught earlier, the secret is very simple and this one is easy to learn and perform. Best yet, it's easy to teach an assistant who can help you.

Effect

A series of cards are laid out on a table and while you're out of the room or off in the distance, where you can't see what's going on, a spectator points to one of the cards to select it. When you come back, your assistant points to various cards. When he points to the selected card, you're able to identify it.

Materials and Requirements

Eight playing cards, one of which is an eight. The suit doesn't matter.

You'll need a table or flat surface for this one.

Secret

The cards are set up in the same manner as the pips on the eight. When your assistant points to the eight, he reveals the position of the selected card. When your assistant points to the selected card, you reveal it.

Preparation

Layout the eight cards in the same manner as the pips on the eight of hearts (which is what we are using for this example), as shown in Figure 11.9.

Figure 11.9
Lay out the cards in the same manner as the pips on the eight of hearts.

Performing the Trick

Tell the spectators to agree on one card as you step away from the table.

When they call you back, your assistant begins to point to each card on the table.

When your assistant points to the eight, he points to the pip that corresponds to the location of the selected card, as shown in Figure 11.10.

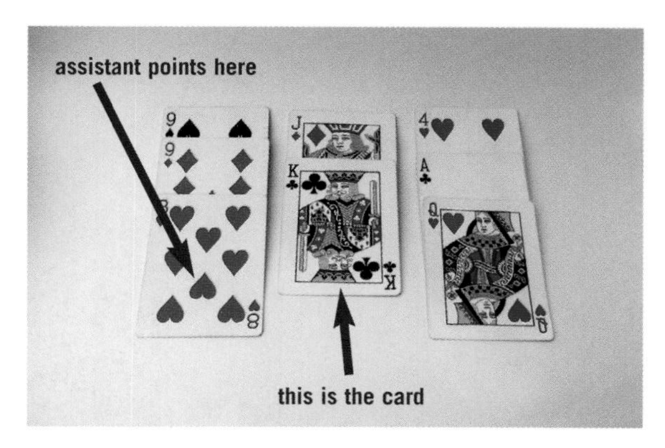

Figure 11.10
Pointing at this pip indicates the card is at this location.

Here's a second example, as shown in Figure 11.11.

Figure 11.11
Pointing at this pip indicates the card is at this location.

Now that you know the card, all you have to do is wait until your assistant points to the card at the designated location and state that it is the chosen card.

Should the eight of hearts be the chosen card, your assistant simply points to its location on the eight. This way, you'll know that eight was the chosen card and can immediately identify it.

Like "Dark Magic," this one can be repeated, but I wouldn't perform it more than twice. If you do choose to repeat the trick, be sure to tell your assistant to mix up the order so it's clear that you're not simply counting to a number. This will keep your spectators guessing.

The Group Coin Vanish

Here's a trick that allows many people to get involved in a trick. In this effect, you vanish a coin.

Effect

You cover a coin with a handkerchief and invite spectators to reach under the handkerchief to verify that the coin is still there. After the last spectator has verified that the coin is indeed under the handkerchief, you whisk the handkerchief away and show that it has vanished. You then show that the coin is now in your pocket.

Materials and Requirements

Two like coins.

A handkerchief, bandana, or cloth napkin.

A secret helper who is in on the trick.

It's best to perform this trick in short sleeves or to have your sleeves rolled up. Spectators will immediately think that you placed the coin in your sleeves under the handkerchief.

Secret

Your helper is the last spectator to verify that the coin is under the handkerchief. When your helper reaches under the handkerchief, he secretly grabs the coin and takes it.

Preparation

Place the second coin in your pocket.

Performing the Trick

Cover the coin with the handkerchief, as shown in Figure 11.12.

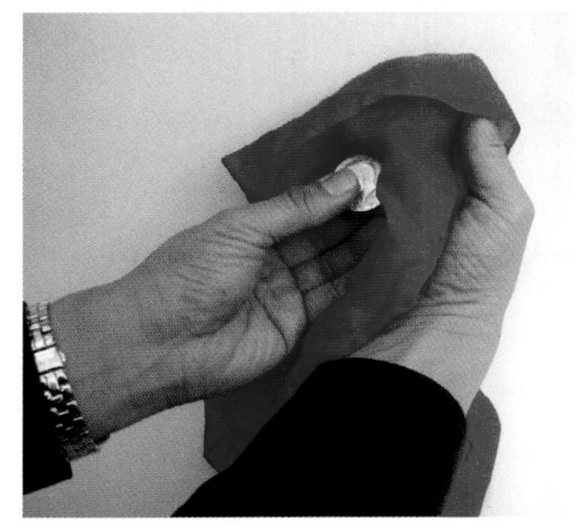

Figure 11.12
Cover the coin with the handkerchief.

Walk around the room and allow different spectators to reach under the handkerchief and verify that the coin is still there.

Go to your secret helper last and invite him to reach under the handkerchief to verify that the coin is there. Your helper secretly grabs the coin and takes it away.

Pretend that the coin is still under the handkerchief. Whisk away the handkerchief to show that the coin mysteriously vanished.

Reach into your pocket to reveal the "vanished" coin.

You can also perform a variation where a coin turns into a different coin or a poker chip. Simply have your helper place the second coin or object in your hand as he removes the first.

A good trick to perform before this one is "Coin Through Handkerchief" and "Coin Through Handkerchief and Finger Ring," both of which are taught in Chapter 7. If you want to perform "Coin Through Finger Ring," you'll need to be using a fine fabric handkerchief, as a cloth napkin will be too thick to twist and thread through a finger ring.

The Impassable Corks

In this impromptu effect, you cause two objects to seemingly pass through each other. All you need for this one are two wine corks. If you don't have wine corks, you can use rolled-up dollar bills or even AA batteries.

Effect

Two wine corks that you're holding appear to pass through each other.

Materials and Requirements

Two wine corks, rolled dollar bills, AA batteries, or Chapsticks.

Secret

The way that you hold the two corks allows them to seemingly pass through each other.

Preparation

This one can be done at any time and there is no preparation.

Performing the Trick

Grip one of the corks in the left hand, as shown in Figure 11.13.

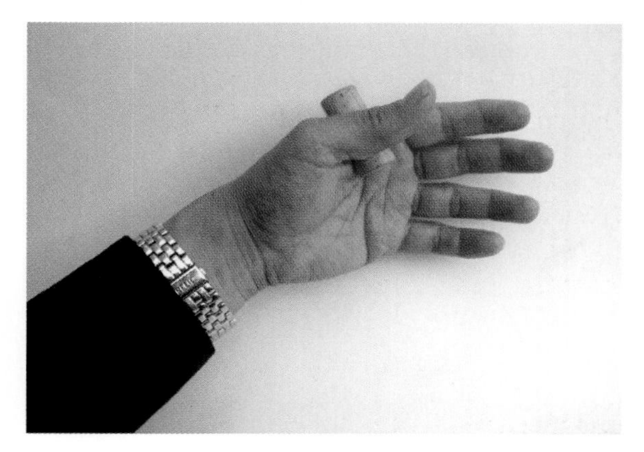

Figure 11.13
Grip the first cork in the left hand.

Grip the second cork in the right hand in a similar manner, as shown in Figure 11.14.

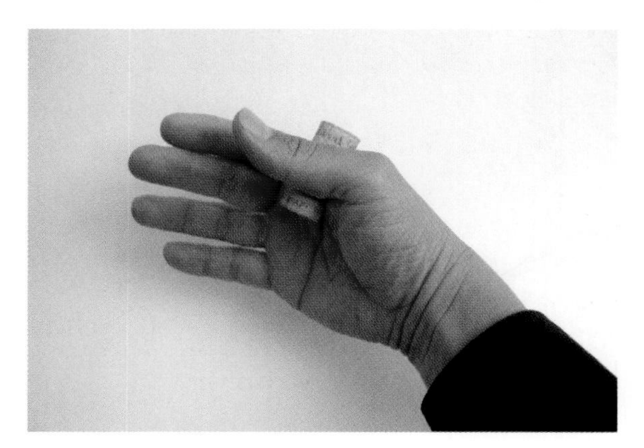

Figure 11.14
Grip the second object in the right hand in the same way.

Place the left thumb on the bottom of the cork and the left first finger on the top of the object, as shown in Figure 11.15. Notice that the corks naturally turn ninety degrees relative to each other.

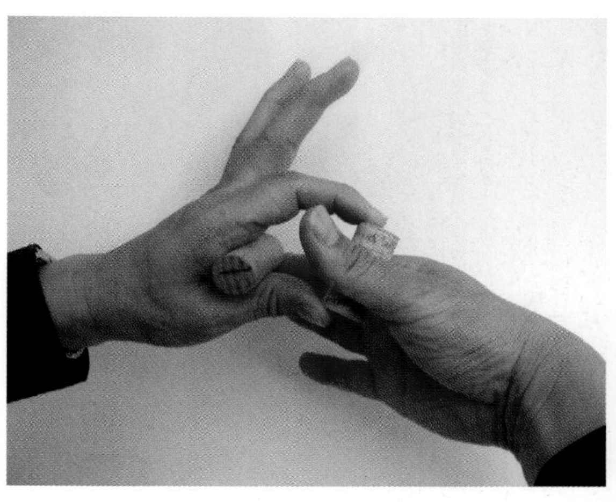

Figure 11.15
Grip the right hand's cork with the left hand.

Here's the crucial part. Your right thumb appears to grip the left hand's cork in the same manner, but you reach your right thumb through the gap formed by your left thumb and first finger to grip the furthest end of the cork, as shown in Figure 11.16.

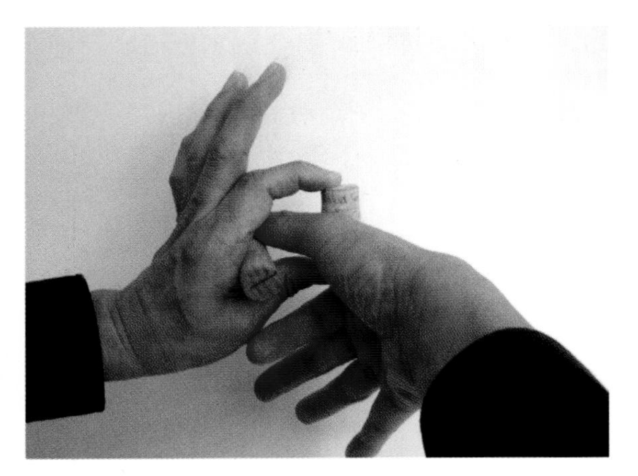

Figure 11.16
Reach your right thumb through the gap formed by your left thumb and first finger to grip the furthest end of the cork.

Your left first finger goes under your right thumb to grip the nearest end of the cork, as shown in Figure 11.17.

Figure 11.17
The right first finger reaches under the left thumb to grip the cork's nearest end.

Reality Check

While it looks as if the corks are locked together by your fingers, the cork that's held by the right hand, because of the odd grip, is actually not locked into the other cork. This position may feel uncomfortable at first, but with practice, this won't be a problem.

At this point, you simply pull your hands away to release the corks. The corks will look as if they passed through each other because you never let go of the corks, as shown in Figure 11.18.

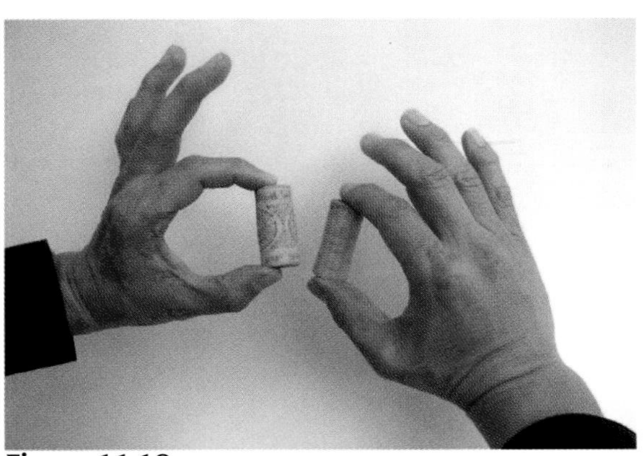

Figure 11.18
The corks will appear to pass through each other.

The Table Routine

This routine is perfect for when you're sitting at a table after dinner. The two-phase routine uses common items and builds to a surprising conclusion.

Effect

In the first phase, you take a coin and cause it to pass through the table from one hand to the other. In the second phase, you tell spectators that you will cause the coin to pass underneath the table under controlled conditions by covering the coin with a salt shaker so you can't touch the coin. To your chagrin, the coin doesn't disappear, but then, to the shock of spectators, the salt shaker vanishes in a stunning fashion.

Materials and Requirements

A coin.

A salt shaker.

A paper napkin.

You'll want to perform this one while seated at a table. The table may have a tablecloth; it doesn't make a difference.

Skills

French drop (Chapter 7).

Finger palm (Chapter 7).

Secret

You use sleight-of-hand to cause the coin to pass through the table. In the second phase, you use strong misdirection to vanish the salt shaker.

Preparation

This one can be done at any time and there is no preparation.

Performing the Trick

Phase one: Coin Through the Table.

Hold the coin to prepare for a French drop, as shown in Figure 11.19.

Figure 11.19
Prepare for the French drop.

Perform the French drop and finger palm the coin in the left hand, as shown in Figure 11.20.

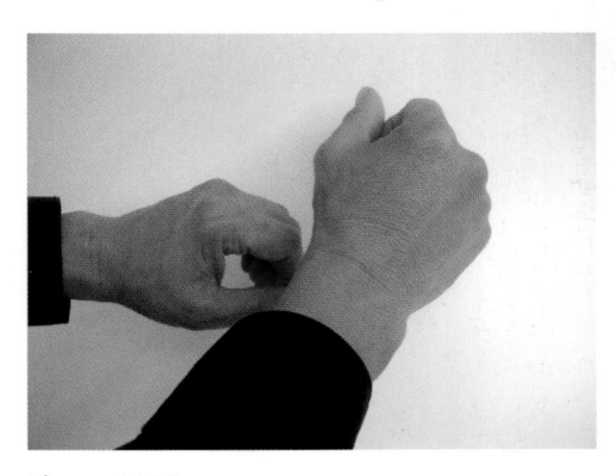

Figure 11.20
Perform the French drop and finger palm the coin.

Tap the top of the table with the closed right hand, as if looking for a "soft spot" and immediately bring the apparently empty left hand under the table.

Slam your right hand down onto the table, as shown in Figure 11.21.

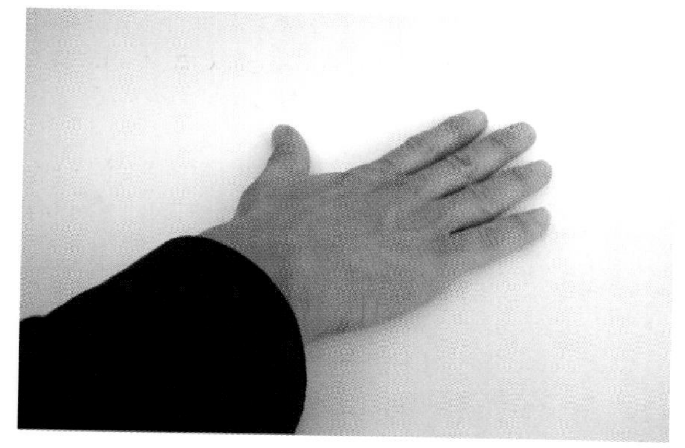

Figure 11.21
Slam your right hand down onto the table.

At the same time that you slam your right hand down, slam your left hand with the coin up onto the bottom of the table. Done right, your spectators will hear the coin slamming down onto the table and going through it. The sound contributes to the illusion.

Turn over your right hand to show that the coin is no longer in your hand and bring out the left hand from under the table to show that it now holds the coin, as shown in Figure 11.22.

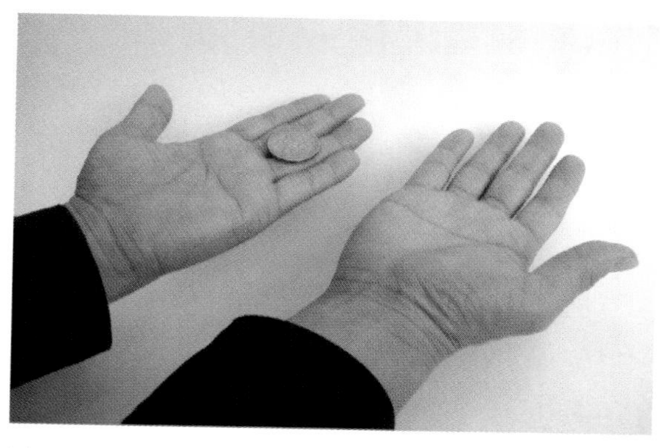

Figure 11.22
Display your hands to show that the coin somehow passed through the table.

Phase 2: The Vanishing Salt Shaker

After performing the first phase, use your left hand to place the coin onto the table. You will want to set the coin down at least a foot from the edge of the table. The distance contributes to the misdirection.

Wrap a napkin around your salt shaker. Make sure that the excess napkin at the bottom of the salt shaker is spread out over the table. You're not actually wrapping the salt shaker; you're tightly covering it with the napkin. You'll need the salt shaker to easily fall out later, which is the secret to the second phase of this routine.

You tell spectators that you will cause the coin to fall through the table once again under controlled conditions. The salt shaker will prevent you from touching the coin. You can also say that you are wrapping the salt shaker to protect the secret. This is shown in Figure 11.23.

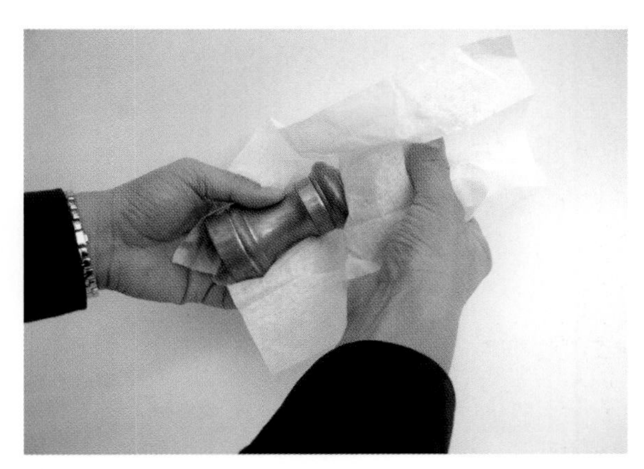

Figure 11.23
Wrap the salt shaker in a napkin. Note that there should be no material under the salt shaker.

Set the wrapped salt shaker down onto the coin, as shown in Figure 11.24.

Figure 11.24
Set the wrapped salt shaker over the coin.

Pause for a moment, snap your finger, and say that even though you couldn't touch the coin, it has now vanished from under the salt shaker. Lift up the wrapped salt shaker to show that the coin has vanished.

When you lift up the salt shaker, quickly bring the napkin and salt shaker to the edge of the table and release your grip on the napkin a bit to allow the salt shaker to fall into your lap, as shown in Figure 11.25.

Figure 11.25
Spectators are looking at the coin as you drop the salt shaker into your lap.

While you are performing this dirty work, the spectators are looking at the coin, which provides strong misdirection. Also, because you have apparently failed at vanishing the coin, this will cause spectators to relax momentarily. The salt shaker is sitting in your lap.

You'll notice that the napkin retains the basic shape of the salt shaker. This is the secret to the trick so don't squeeze or crush the napkin. Just bring back the empty napkin and gently place it over the coin and rest it on top, as if the salt shaker is still in it. During this offbeat moment, spectators will assume that the salt shaker is still under the napkin. This is shown in Figure 11.26.

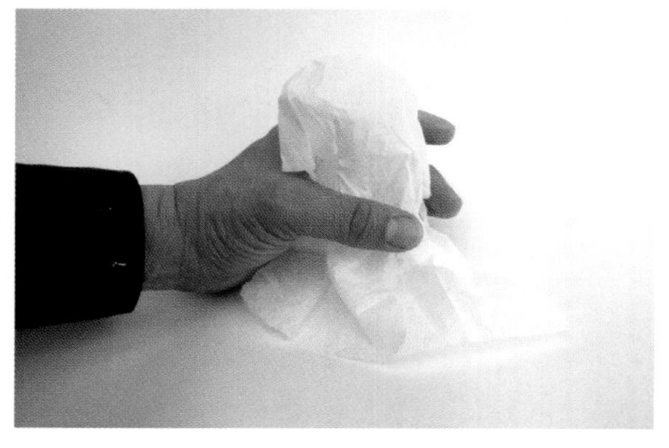

Figure 11.26

Rest the empty napkin shell over the coin. It will look as if the salt shaker is still wrapped by the napkin.

Say that you forgot that this trick doesn't involve vanishing the coin, but the salt shaker. Take your hand and slam the napkin down onto the table to completely flatten it. To the spectators, the salt shaker has mysteriously vanished in stunning fashion. This is shown in Figure 11.27.

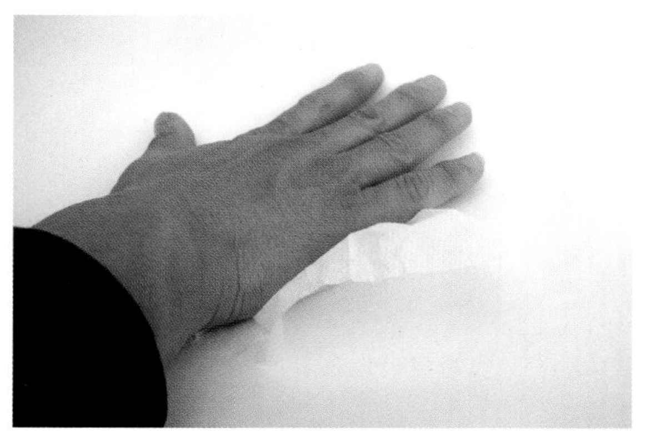

Figure 11.27

You slam your hand down onto the napkin and the salt shaker disappears.

You have two choices afterwards. You can bring out the salt shaker from under the table to show that it passed through the table as the coin did. But perhaps the better thing to do is quietly keep the salt shaker and perform another trick to take spectators' attention away from the whereabouts of the salt shaker. Or you can allow the party to return to normal and later, quietly bring back out the salt shaker onto the table when no one is looking.

I personally think that the trick is stronger as a vanish than as a salt shaker through table effect.

A good coin trick to perform before this effect is "Coin Routine: Elbow to Elbow" (Chapter 7). You may also consider "The Ten-Count Coin Trick" (Chapter 7).

Thirst for Knowledge

Here's another secret code that is not only effective at parties, but works in restaurants and bars. Once again, you're working with someone in the know and the secret is easy and difficult to detect.

Effect

A series of small objects are placed onto a table. You walk out of range and a group of spectators decides on an object. You rejoin the group and hold your hand over the different items, one at a time. When your hand is over the designated item, you're able to tell spectators that you've found the object.

Secret

Yes, it's a visual code and the secret is in the name.

Preparation

An assistant who can secretly communicate with you.

Performing the Trick

Gather a series of items on a table and ask spectators to secretly select one as you step away.

When you come back, you say nothing and look down at the objects on the table. You hold your hand over the various objects, one at a time, as if to sense a vibration. At no time do you seem to be looking for a signal from someone and you remain focused on the items at hand.

When your hand passes over the selected object, your assistant simply takes a sip of his or her drink. You now know the object and can immediately identify it or come back to it later. Try it out. You can repeat this one and should have no problem fooling everyone again.

Other Tricks for Parties

Many of the card tricks in Chapters 2 and 5 will work well at parties. In particular, we like "Picture Yourself As a Magician" (Chapter 5), "Do As I Do" (Chapter 5), "Four Aces" (Chapter 5), "Aces High" (Chapter 2), and "The Searchers" (Chapter 2).

Next Steps

Now you can move onto Chapter 12, which will teach you how to levitate. If you're ready to literally reach for the sky, you'll learn how.

Mr.B's
Closing
Sale

Levitation

IT'S PERHAPS THE ULTIMATE MAGIC TRICK. Somehow, you make yourself lighter than air and break the binding laws of gravity. It's a classic dream from days of old. And today, when compared to finding a selected card or causing a coin to vanish, you'll definitely receive some attention when you lift yourself off the ground.

Flying High

STAGE MAGICIANS HAVE LONG performed levitations, usually causing an assistant to somehow float in midair or rise from the stage floor. In these illusions, there are numerous "convincers" that magicians add to the mix. The magician passes a solid ring over the floating assistant or swings a pole or sword over and under to prove that there are no wires or means of support.

While illusionists perform levitations and make them look convincing, audiences inherently know that it's some kind of trick. Few truly believe that someone is indeed floating on stage and many convince themselves to believe that the secret is somehow tied to the controlled conditions of a stage in a theater, whether it is or not.

But what if you could levitate while standing on a sidewalk? Out of the theater, surely, spectators would be able to see a crane or some supporting structure. Yet, as you'll see, there are ways to make it appear as if you are floating off the ground. You won't be reaching any significant heights as these methods of "self-levitation" have limits. But you can astound and mystify your spectators.

We offer three techniques for levitating yourself just about anytime and anywhere.

Get ready to reach for the sky.

The Balducci Levitation

This levitation makes it look as if you are floating off the ground between four and six inches, as shown in Figure 12.1. You can perform this one anywhere, but you have to be careful about angles and your distance from spectators. And you can only perform this for a small group of people, probably no more than five who are tightly grouped together. By the way, the levitation is named for Ed Balducci who was the first to publish it, but the originator of the effect is unknown.

Figure 12.1
The Balducci levitation.

Secret

You are actually lifting yourself by standing on the toes of one foot. Because of the way that you stand in relation to spectators, your front foot covers the lifting action of the other foot. With this levitation, angles and positioning are everything.

Preparation

None.

Materials

None.

Performing the Trick

You'll have to do a lot of experimenting and practice to make this one work. The idea is that you stand a certain distance from a small group of spectators, say six to eight feet, and you turn your back to spectators and then turn approximately 45 degrees towards the foot that you will lift off the ground. Figure 12.2 gives you a closer view of the position of the feet; this is what spectators see.

Figure 12.2

This is what spectators see. The distance and angles are important.

When you lift on your toes, the spectators' sight lines prevent them from seeing your lifting foot—the closer foot blocks their view. The spectators see the back of the heel of the lifting foot, but the front foot blocks the toe that is in contact with the ground. Figure 12.3 shows what is really happening in an exposed view.

Note how the foot that is lifted off the ground remains at the same angle as if it were still standing on the ground. This adds to the effect.

Figure 12.3

Here's the exposed view. Note how the "lifted" foot remains at the same angle as if it were still standing on the ground.

Figure 12.4 shows the other side.

You'll need to practice with a helper who can provide feedback on angles. When you can consistently place yourself relative to your helper and perform the levitation, and he can't see your lifting foot, you're ready to try this with real audiences.

Figure 12.4
The view from the other side.

A tip: the baggier your pants, the better, as this will help hide the lifting foot. Also, the straighter the edge along the back of your heel, the stronger the illusion. Shoes with heels that curve underneath your foot can make the lifting foot appear to be doing exactly what it's doing: carrying you upwards by standing on your toes.

Try to get a momentary lift and then immediately come down. When spectators ask you to perform the levitation again, you can say that you only have one of these a day and can't always get "lift."

Obviously, anyone out of your spectators' sight line will see exactly what you are doing, which will expose the trick. You can only perform this levitation within its strict limitations. But despite this, you can do this one at any time when the viewing conditions are right.

The Jacket Levitation

In this method, you levitate under the cover of your jacket. This one requires some preparation.

Effect

You take off your jacket and momentarily drape it over your shoes. With your shoes showing, spectators see you and your shoes rise from the ground. You can perform this one anywhere and at anytime, but spectators have to be directly in front of you. The use of the jacket gives you far wider angles than the Balducci levitation. This is shown in Figure 12.5.

Figure 12.5
The jacket levitation.

Secret

As in the Balducci levitation, you are standing on the toes of one of your feet. Your shoes are kept together with Velcro.

Materials

A package of Velcro tape or squares of sticky Velcro fasteners. We used the squares of Velcro.

A jacket.

Slip-on shoes that you can easily take off without the use of your hands. You can also perform this with a loosely tied shoe. Dark shoes may help to hide the Velcro, depending on the kind of tape that you are using.

Preparation

Stick the hook side of the Velcro tape or fasteners to the rightmost side of your left shoe at the wide points of the shoe near the ball of the foot and the heel. Stick two fasteners of the fabric side of the Velcro to the leftmost side of your right shoe at the same points.

The Velcro fasteners are shown in Figure 12.6. For demonstration purposes, we used white fasteners on black shoes. If you want to perform this trick, you will want to use tape that matches your shoes. Yes, this looks ridiculous, but sometimes, you do have to suffer for your art.

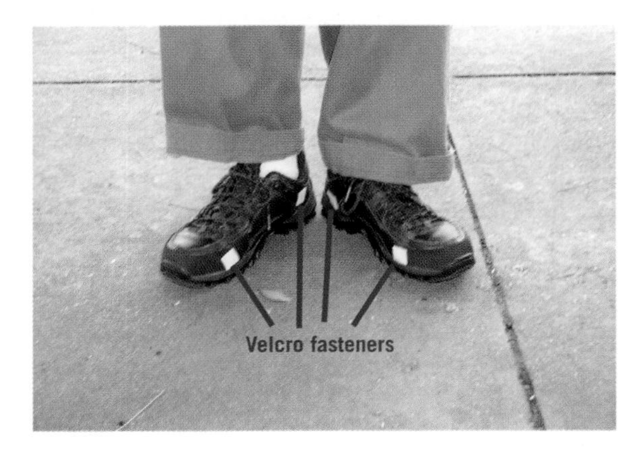

Figure 12.6
Stick Velcro fasteners to the wide parts of your shoes on the inside edges.

You'll have to experiment with the amount of Velcro fastener that you'll need, as well as with the adhesive. We found that two squares of Velcro were sufficient for holding a lightweight shoe, but the adhesive had to be enhanced with additional glue.

After some experimenting with Velcro and glue, when you place your shoes together, they will bind together. And you can also release your shoes by simply twisting your feet. The more Velcro you use, the more securely the second shoe will attach to the first. However, the more Velcro you use, the more sound you'll produce when you detach your shoes. You want to strike a good balance between security and sound.

Another thought: you can purchase sheets of Velcro from fabric and craft stores that you can cut to fit your shoes. In particular, if your shoes have a particular design, you can incorporate the Velcro fabric into the design to better mask it.

> **Warning**
>
> Be careful when walking with these shoes that you don't inadvertently bring your shoes together and trip.

Performing the Trick

Before you perform the trick, be sure that your shoelaces are loose so you can readily slip off your shoe.

Make sure that spectators are standing directly in front of you about five feet away. You want their sight line to look slightly down on your feet.

Take off your jacket and stand with your feet together. Rest the bottom edge of your jacket on the ground just in front of your shoes. Under cover of your jacket, allow your right shoe to adhere to your left shoe. This is shown in Figure 12.7.

Figure 12.7
Drape the jacket on the ground in front of your shoes and allow your shoes to adhere together.

Secretly slip your right foot from its shoe and rest it about a foot back, as shown in Figure 12.8. You're ready to levitate.

Figure 12.8
Slip your right foot out of its shoe and set it back a bit under cover of the jacket.

Bring your jacket back and up and expose the tips of your shoes, as shown in Figure 12.9.

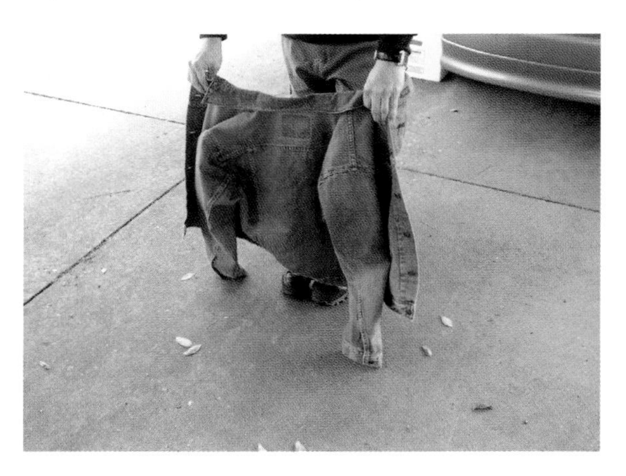

Figure 12.9
Bring your jacket back and show the tips of your shoes.

Reality Check

The spectators see the tips of two shoes from under your jacket and assume that these are your two feet. Your real right foot is actually behind your jacket ready to lift you up.

Balance yourself on your right foot but keep your left foot extended at its location. Tell spectators to watch your feet and lift up your left foot with its accompanying shoe, and at the same time, lift your jacket to move with your shoes. To spectators it will appear that you are lifting off the ground. This is shown in Figure 12.10.

Figure 12.11 shows how it appears in the front.

Figure 12.10
Lift your left foot and its accompanying shoe along with the jacket in one coordinated motion.

Figure 12.11
This is how it appears to spectators.

Slowly lower yourself and bring down your left foot and its accompanying right shoe.

Reverse the setup. Momentarily bring your jacket forward so it's touching the ground and hiding your shoes. Secretly place your right foot back into its shoe and separate the right shoe from the left shoe. With practice, this should take about a second.

Lift up your jacket and put it back on.

This effect will cause your feet to rise but won't cause your body to appear to rise. This is why it's important to direct the spectators' attention to your feet. In theory, if they're watching your feet rise, they won't realize that your body didn't rise with it. You can also experiment with lowering your body a bit by bending your knees as you adjust the jacket over your feet. When you levitate, you lift your shoe and straighten your legs to make it appear as if your body rose with your feet.

This effect has two variations. In the first, when you lift your left shoe and jacket, you can also lift your entire body onto the toes of your lifting foot. This will make your body appear to rise at the same time. Of course, standing on your toes while extending your other foot is somewhat tricky and will take some practice.

In the second variation, you can appear to turn while in the air. For this, you simply rotate your left foot and its accompanying shoe in one direction and lean your body in the other. Also, make sure that you adjust your jacket accordingly. This is shown in Figure 12.12.

Figure 12.12
By tilting your body, you can appear to rotate while floating in the air.

The Box Levitation

Here's a levitation that you can perform almost anywhere, although you'll want some distance between yourself and spectators. The use of a box effectively masks the secret. There's some work in creating the gimmicked box, but you can find the materials you need in office supply and hardware stores and you'll also need access to a digital camera, computer, and color printer.

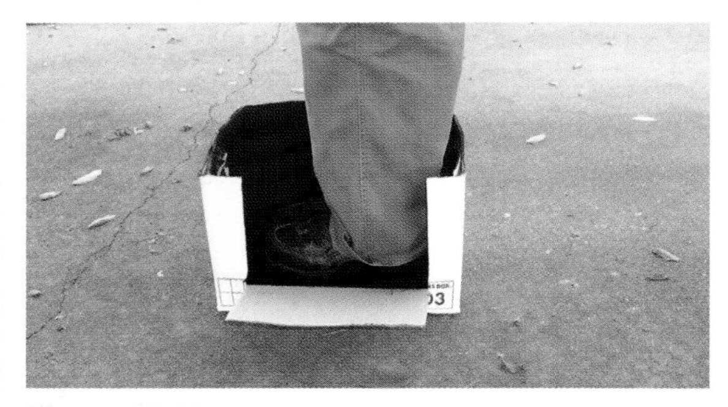

Figure 12.13
The Box levitation.

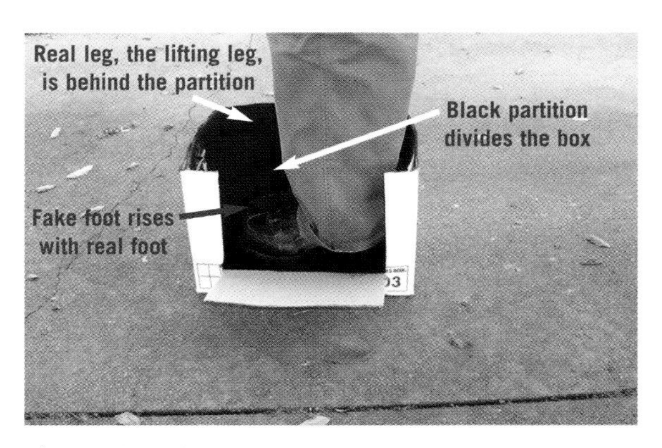

Figure 12.14
Spectators see your foot and a fake foot rise in the box. Your foot, which you are using to lift yourself up, is hidden in a secret compartment.

Secret

As in the Balducci and jacket levitations, you float by lifting yourself up on the toe of one foot. However, this time, your support foot is hidden in a secret compartment in the box. Spectators see your other foot and a fake foot slowly rising with your body, as shown in Figures 12.14.

Most of the work here is in creating the box. The box levitation gives you better angles than the Balducci levitation by effectively masking off bad ones. The box basically provides blinders that protect the levitation's secret. The fake foot is attached to the box using rubber bands and rises with your foot.

Materials

A cardboard box, approximately, 15" x 12" x 10.5". We used an office storage box purchased from an office "big box" store. The box will also need a lid that fits on top and is not connected to the box.

Two sheets of heavy cardboard, poster size.

Black paint (flat).

A box cutter or scissors.

Glue—glue sticks are fine.

Duct tape.

A rubber band.

Some paper clips. The stronger the better.

A binder clip.

A spring—You can get this from your hardware store. Get a soft spring about two inches long that compresses and bounces back. We got our spring for about two bucks.

Preparing the Fake Foot

Put on the pants and shoes that you will be wearing when performing the levitation. Rest your right foot on the ground and pull your left leg back and have a friend shoot a picture of the left side of your right foot and lower leg. If you want to perform this using the opposite leg, you'll be rising on your left leg and displaying your right leg, simply reverse the directions.

Upload your picture to a computer. You'll need a program that can expand the photo so it fills a page that's in landscape (wide) mode. Programs such as Adobe Photoshop and other image editors, as well as desktop publishing programs will allow you to do this. If you have a large foot that won't easily fit on a single page, you may have to break the image into two parts and print them and tape them together later.

At this point, lighting is crucial. If you will be performing under dark conditions, you may have to darken the image somewhat. Print out the picture of your shoe and compare it against your real one. You'll want a picture that matches your real shoe as close as possible under the lighting in which you will perform. If you have the software tools, "cut out" the image of your shoe and pants and place it on a white background to save on ink when you print your image. Figure 12.15 shows a fake foot.

You can also use a traditional film camera and have the film developed and purchase an 8"x10" or 10"x14" enlargement of your foot. However, you won't have the ability to adjust the lighting and contrast of the foot to match your lighting conditions.

Figure 12.15
Take and print out an image of your right foot.

With the color image in hand, glue the image to some light cardboard and cut out the image of your foot and lower leg. Make sure that the bottom of the shoe is straight. You'll end up with a gimmick of your leg and foot.

Unbend a paper clip and using the duct tape, attach it to the back of the fake foot. Later, you'll be attaching a rubber band to this clip.

Figure 12.16
Attach an unbent paper clip to the back of the fake foot.

You'll also need a platform that is securely taped to the bottom of the fake foot. For this, you can use unbent paper clips that are at 90-degree angles. Your real foot rests against these and controls the position of the fake foot. If you like, you can use L-shaped angle braces that won't tend to bend the way that paper clips do.

Preparing the Box

Figure 12.17 shows our original ungimmicked box.

Figure 12.17

The ungimmicked box.

Cut two slits down from the top, front edge of the long side of the box so you end up with a rectangular section. This creates a "door" that you can bend to hide your setup and then display your levitation. Paint the inside of the box black. This is shown in Figure 12.18.

Figure 12.18

Cut out a rectangular portion on the long side of the box and paint the inside of the box black. Note that we left a portion of the box unpainted to help clarify where the "partition" is inserted later.

Using a sheet of cardboard, cut out a rectangular piece that fits into your box lengthwise. Paint one side of this cardboard black and using the duct tape, tape it into the box, as shown in Figure 12.19. Make sure that you only use the tape on the back, non-black side of the cardboard.

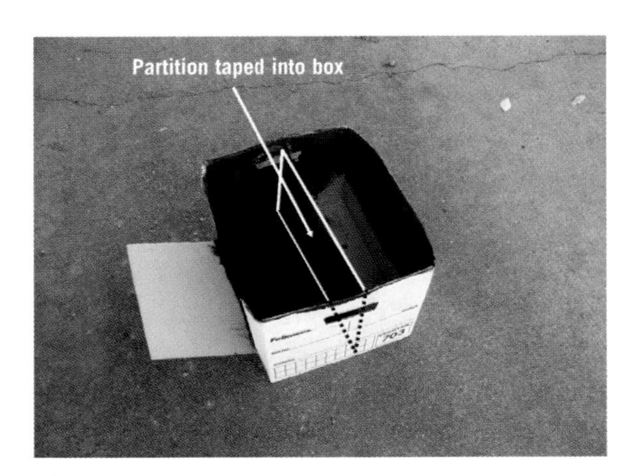

Figure 12.19

Cut a sheet of cardboard and paint it black and tape it into the box.

Stand in the box with your right foot to the right of the cardboard and your left foot on the other side. Make sure that you have room to move your feet and it's not too cramped. Test the box by standing on the toe of your right foot.

If you like, you can use another sheet of cardboard and cut a rectangular piece that will form a door to the box. You can tape this cardboard to the box and use the tape as a hinge. You can also use another piece of tape as a latch to keep the door shut when you need to. Of, if you prefer, you can use Velcro pieces to secure your door. You'll probably want to take the time to do this if you'll be using the box a lot.

Attach a binder clip to the top of the "partition" and then use a rubber band to connect the binder clip to the paper clip that is on the back of the fake foot. The binder clip should reside in a location that is covered by your left leg when you're standing in the box. You'll have to experiment a bit with the placement of the gimmick and the size and tension of the rubber band to ensure that the fake foot rises properly with you when you rise.

If the gimmick is not sitting right or does not move with your left leg when you are rising, go back and adjust the binder clip and the rubber band. The fake foot should reside side by side with your real foot and move with it and maintain its orientation. Obtain a new rubber band if you need to. The fake foot is shown in Figure 12.20.

Figure 12.20
The fake foot set up in the box.

Use a third unbent paper clip that's bent at a 90-degree angle and attach it to the spring. Attach the other end of the spring to the bottom of the box. This is all shown in Figure 12.21.

Figure 12.21
This picture shows the lifting mechanism.

Before you perform the trick, the cardboard shoe and lower leg image should simply hang in the box, covered by your "door" and the lid of the box. You're ready to perform the trick.

Performing the Trick

Position spectators about five feet away from the box on the door side. You want spectators to be looking slightly down.

Lift up the lid and use it to shield the spectators' view of the inside of the box. Most importantly, you don't want spectators to see the top of the fake leg. Step into the box with your right foot to the right of the partition and your left foot to the left of the partition. Step on the platform that's connected to the fake foot and drop your weight onto your left foot. Make sure that the fake foot goes down to the floor of the box with your left foot. Place the lid down in a location where you can easily grab it again.

Bend over and open the front door of the box so spectators can look in. Your leg in the front compartment hides the hardware.

> **Reality Check**
>
> Spectators see two legs and shoes, but they're really seeing your left leg and an image of your right shoe and leg. Your right leg is in the secret compartment ready to lift your body.

Stand up straight and slowly lift up on your right toe. Keep your left foot level as it rises and the image of the shoe and leg will rise with it. Momentarily stay in the topmost position, as shown in Figure 12.22.

Figure 12.22
Levitate by standing on your right toes and keeping your left foot level as you rise.

Drop back down until you're standing on your right foot and your left foot is back down on the floor of the box.

Bend over and close the front door. Pick up the lid and use it to shield the audience's view of the inside of the box from the top as you step out of the box. Place the lid on top of the box.

Constructing the box may seem like a lot of work for an effect that only lasts a few seconds. But, you appeared to briefly float up into the air. Because of the nature of this effect, it has to be short and sweet.

You want to quickly amaze spectators and then come back down and close the door and put back on the lid before they have time to think about what they saw and analyze it. Let them imagine some complex hydraulic system in the blackness of the box that causes you to mysteriously lift up. If you're performing this levitation in a show, you can casually bring out the box and after the effect, easily lift up the box and set it aside, which will suggest to spectators that there is no heavy equipment inside.

Another consideration: you can obtain more "lift" if you insert a block of wood in the hidden compartment. This block goes under your right toe. With your right toe resting on the block and right heel resting on the floor of the box, you can easily step up onto the block to more easily raise your body. This is something with which you can experiment.

Next Steps

In the next chapter, you'll be learning to perform cutting-edge magic using electronic gizmos such as iPods and PlayStation Portables.

MENU

13

Electronic Magic with iPods and More

OUR MODERN-DAY TECHNOLOGY is practically magic in itself. And since we can all carry around devices such an iPod, Nintendo DS, and PSP game system, why not perform some magic tricks that rely on the capabilities of these devices.

Virtual Magic

WHAT I'VE TRIED TO DO in this chapter is to create a series of tricks where the magic occurs on the screen of your electronic device. The tricks here work with video that I provide that plays on the screen of your electronic device. For explanation purposes, we will use a video iPod throughout, but the instructions also apply to a Sony PSP. Just upload and use the PSP-related video files.

With these electronic magic tricks, you can reveal a chosen card or an object that was selected by your spectator. All you have to do is set up the trick, force the card or the object, as you've learned in earlier chapters, and then let the video make the stunning revelation.

I have also developed tricks where it looks as if you are inserting your finger into the screen, or laying the iPod across your hand, and the LCD acts as a window that spectators can see right through. I love these kinds of magic tricks because they effectively blur the line between virtual electronic worlds and reality.

By the way, the electronic files that are related to each trick are provided on the enclosed DVD. While the DVD acts like a movie DVD when you insert it into a DVD player, it also acts as a data disc when it's inserted into the DVD drive of a computer. Simply copy the files to your computer and then upload them to your iPod so you can use them.

I hope that you have fun with these cutting-edge effects that I have developed.

Card Detector

This first effect makes use of the fact that you can play video on your electronic device. It's a card revelation.

Effect

A spectator selects a card. You fail to find the card and then pull out your trusty iPod. Your iPod displays a series of playing cards and slowly moves around and then zooms into a single card. Of course, it's the spectator's selected card. The first phase is shown in Figure 13.1.

Figure 13.1
Your iPod shows a series of cards.

Secret

You force the jack of spades, you do a little acting, and the video does the rest and reveals the card for you. The second phase, the revelation, is shown in Figure 13.2.

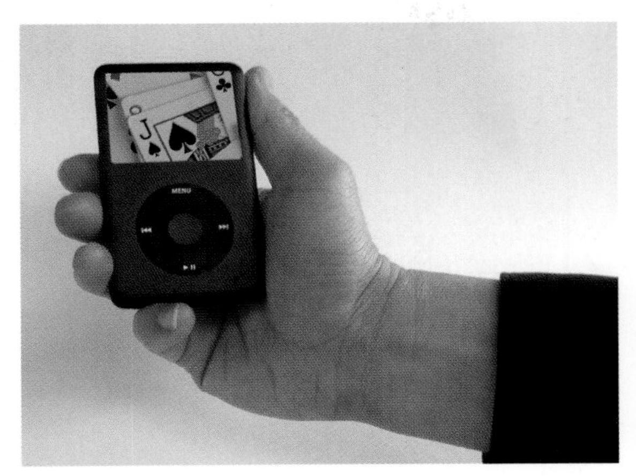

Figure 13.2
Your iPod zooms in and stays on the jack of spades.

Figure 13.3
The spectators see the palm of your hand through the iPod's LCD screen.

Materials

A deck of cards.

Video File

For this trick, you'll need to upload and use video file **carddetector.mp4**.

Performing the Trick

You can use your choice of any card force including the cross-cut, cut deeper, or Hindu force. You can read about these in Chapter 4.

Card Transporter

Effect

In this card effect, you rest your iPod on your hand and the spectators seem to see your palm through the iPod's LCD screen. This is shown in Figure 13.3.

Slowly, something materializes on your hand, and it's the spectator's selected card, the five of diamonds. This is shown in Figure 13.4.

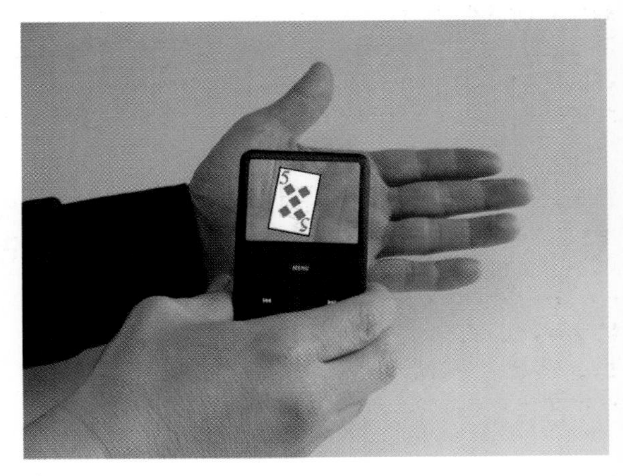

Figure 13.4
The five of diamonds gradually appears on the palm of your hand.

The five of diamonds then disappears as mysteriously as it appeared. And then you remove the iPod from your hand.

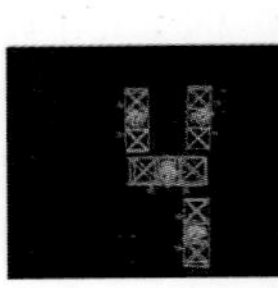

Secret

Once again, you force the five of diamonds using any force that you wish. Just use one of the forces found in Chapter 4.

Materials

A deck of cards.

Video File

For this trick, you'll need to upload and use video file **cardtransporter.mp4**.

Performing the Trick

Force the five of diamonds.

Begin the video and place the iPod on the palm of your left hand.

Watch the video cause the selected card to appear on top of your hand.

The Conversation

Effect

Six objects are placed on the table and you and a spectator gradually eliminate objects. You tell your spectator that your iPod knows what object was selected. You play a video and hold a conversation with your iPod and it reveals the chosen object.

Secret

You force a single object, a wrist watch, which the video reveals.

Skills

To force the wrist watch, use the force that you learned in "Your Destiny 2," which is in Chapter 8.

Materials

Six objects.

Video File

For this trick, you'll need to upload and use video file **conversation.mp4**.

Performing the Trick

Gather six objects and make sure one of them is a wrist watch. It helps if you allow spectators to contribute various objects from around the room or area. As long as one of the objects is a wrist watch, you can perform this trick. And the more unusual the other objects, the better.

> **Hint**
>
> After the trick, spectators will think that you had a video for every object that was originally on the table and simply played the video that related to the final object. Of course, if spectators are contributing objects for the trick, they can literally bring anything to the table. This strengthens the effect.

Perform the force as explained in "Your Destiny 2" in Chapter 8, which will leave the watch on the table. Make sure that you're holding the iPod in your left hand.

Play the video and "converse" with the iPod. This is shown in Figure 13.5.

Figure 13.5
You carry on a virtual conversation with your iPod.

The video will ultimately reveal "Watch." This is shown in Figure 13.6.

Figure 13.6
The iPod reveals "Watch."

Here's the script:

iPod: [Eye appears, opens, and closes]

You: Hello, wake up.

iPod: [Eye appears, opens, and looks around]

You: Over here.

iPod: [Eye appears and looks to its right towards you]

You: Hello.

iPod: Hello.

You: How are you?

iPod: I'm doing fine.

You: We're performing a trick here. Can you help us?

iPod: I'll see what I can do.

You: My friend here just chose an object. Can you tell me what it is?

iPod: [Eye appears and begins to look around]
[If you like, you can cover the "eye" so it can't "see"]

You: No, you can't just look, you have to figure it out.

iPod: [Eye shuts and screen clears] Hmmmmmm.....

iPod: Watch. [Eye appears and then shuts]

It's important that you practice the dialogue so you get the script and timing down. Acting skills will help a lot with this one. You not only have to perform the trick, but you have to treat the iPod as if it's a real personality or person who is thinking and responding.

Virtual Money

This is a slick coin production that will use some of the skills you learned in Chapter 7.

Effect

You place your iPod onto the palm of your hand and a quarter begins to slowly fade into view. At the end, you lightly shake your iPod and a quarter falls from the screen, seemingly from the iPod onto the table.

Video File

For this trick, you'll need to upload and use video file **virtualmoney.mp4**.

Materials

A quarter.

Preparation

Hold the quarter in your left hand.

Performing the Trick

Begin the video.

Keep your left hand curled down with the quarter in it. Bring up the iPod and grasp it with the thumb of your left hand. Turn your left hand and open it and at the same time, place the iPod on top of the quarter and rest the iPod against your hand. Done correctly, your spectators should not see the quarter.

At this point, you have the iPod resting against your palm with a quarter behind it.

Allow the spectators to view the picture of your palm and then watch the quarter slowly materialize on your hand. This is shown in Figure 13.7.

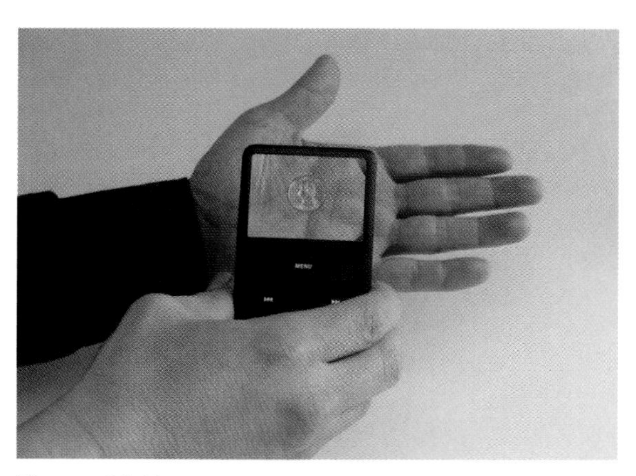

Figure 13.7
A quarter appears to materialize on your hand.

You'll see a flash of light come from the quarter. At this point, lightly shake the iPod and you'll see the quarter begin to fall down.

As the quarter falls down, drop the quarter that's held in your hand. Done correctly, it will appear that the quarter materialized on the screen of the iPod's LCD and then fell out of the screen and onto the table. This is shown in Figure 13.8.

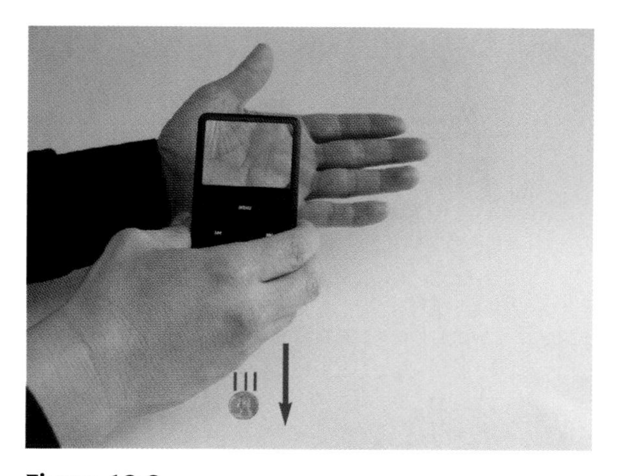

Figure 13.8
The quarter appears to fall from the LCD screen onto the table.

Reach In

Effect

You reveal a spectator's card on your iPod but it's the wrong one. Not to worry, you simply reach into the screen with your finger and push the top card away to reveal the spectator's card.

Secret

You force the seven of clubs using your favorite force.

Video File

For this trick, you'll need to upload and use video file **reachin.mp4**.

Performing the Trick

Force the seven of clubs.

Hold the iPod in your left hand and play the reachin video. Spectators will have to look over your shoulder to view the iPod.

When the LCD displays a jack of hearts, allow the spectator to tell you that it's the wrong card.

You'll see a flash of light from the jack of hearts and you can then reach your finger in from the right side of the screen. You'll see a finger appear on the screen. This is shown in Figure 13.9.

Simply mimic the movements of the on-screen finger as if it were your own. The on-screen finger will push away the jack of hearts to reveal the seven of clubs underneath it. This is shown in Figure 13.10.

Figure 13.9
The LCD screen reveals the wrong card and you appear to reach your finger into the screen.

Figure 13.10
Your finger appears in the iPod's LCD screen and pushes away the top card to reveal the right one.

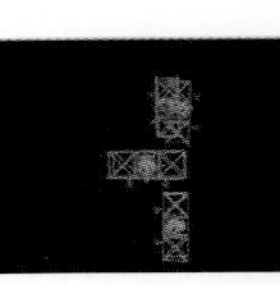

Magic with Your Nintendo DS

NINTENDO RECENTLY INTRODUCED "Master of Illusion," a game that allows you to use a Nintendo DS to perform magic tricks. The game comes with a marked deck that allows you to perform card tricks. It's an intriguing and creative product.

Because this game is so popular, you may find that many kids already know the tricks. And their magic-weary parents have probably long tired of watching the tricks, even if they don't know how they are done. But you can bet that many parents have been learning as well.

While I was generally impressed with "Master of Illusion," the tricks in the DS title are probably too well known for a serious magician to perform. Despite this, I have some suggestions for using the program that will allow you to use your own deck of regular playing cards, which will be a step up from Nintendo's cards.

Many of the tricks rely on the fact that you know a selected card by reading some markings on the back. And as the trick plays on the Nintendo DS, you input a series of commands that secretly tell the DS what card to reveal at the end.

The strength of using Nintendo's marked deck of cards is that the choice of a card is a free one. The downside of using this deck is that it's not a standard one and will always be viewed with suspicion. By using your own deck of Bicycle or other popular brand card, you can still perform the tricks, and can even borrow a deck of cards from spectators.

The key is to already know the card that a spectator selects or find out after it's been chosen and returned to the deck. For this, you already know a series of card forces that you learned in Chapter 4, and you can use any one of them: deeper cut, cross-cut, Hindu, and more. The card can be returned to the deck and mixed, and then you can go about revealing the card.

The second method involves having a card returned to the deck and then finding out its identity by using a method such as a key card, which you learned in Chapter 2. And with objects and other revelations, you can use the forces that we explained in Chapter 8. In any event, you can use your magic skills to take Nintendo's "Master of Illusion" even further.

Next Steps

In the next chapter, we'll talk about ways to perform your magic and entertain audiences. If you've stayed with the book this far, you have the tools and the foundation to move to the next step.

7 ◆

14

The Art of Performing Magic

THERE'S A HUGE DIFFERENCE between knowing and understanding magic secrets and going out there and performing and entertaining with magic. While fooling spectators is a big part of magic, it's equally, or even more important, to present magic in a positive manner that allows spectators to have fun and be entertained.

No one really cares much for a showoff or know-it-all. And even fewer like an individual who implies that he's smarter than everyone else. Yet for some reason, there are many amateur magicians and even some pros that make this devastating mistake.

If you make your magic entertaining, this in itself acts as misdirection that keeps spectators from thinking about "catching" you and figuring out your secrets. Your goal as a magician is to polish your magic so it's strong and hard to figure out, and make it as entertaining as possible, whether you use comedy, storytelling, or music. They go hand in hand.

Presentation

SOME MAGICIANS FEEL that it's best to write down in detail everything that they are going to say and then memorize it. There is a lot of merit to this. By writing down your patter and coordinating it with your moves and routine, you can sort out potential trouble spots when you're performing a crucial move and say something to avert your audience's attention.

I prefer to outline what I'm going to say and then talk through the routine. Because I sound different when I write something than when I say it, I think that this allows me to sound more natural and more like myself. I also outline the funny lines and then practice performing the routine with the dialogue.

You may be wondering where you can get ideas for routines. I like to create a theme for each routine that I perform. Admittedly, I don't always succeed at this. And in some of my standard routines, I simply talk about what I'm doing. But I think that a theme that's based on some scientific principle, old belief, historical incident, haunted object, ESP, teleportation, and more, will always add to your effects and make them more interesting. This is your opportunity to inject your personality into your routines, as well as your interests.

Practice and perfect your routines until you can do them well. You deserve to be at your best and leave a great impression with audiences. Your audiences who invest their time into your performance deserve great entertainment. And the art of magic deserves to be presented in the best manner possible.

Polishing Your Magic

WHEN YOU ARE LEARNING BASIC moves, this is a great time to practice in a mirror. Here, you can slowly perform the moves, learn how to hold your hand and body in a natural manner, and execute each move so it's as smooth as possible.

At this stage, you'll want to practice until you can perform the moves and routines without thinking about them. If you can flawlessly execute the mechanics of a trick while thinking about what you're going to have for dinner, this is a good sign. It means that the moves are ingrained in your head.

©istockphoto.com/Mirela Schenk

After you feel comfortable with the moves, the next stage is to use video so you can see how audiences see you. When you perform in a mirror, you're always seeing yourself from one perspective and angle. But in the real world, audiences may be at your sides, or if you're standing on a stage, their eyes may be slightly below you.

Record yourself and watch yourself on video and you'll learn a lot about how you move and how you present your magic, and this will give you feedback that allows you to make adjustments and improve your performance.

Dealing with Nervousness

MANY BEGINNING MAGICIANS have trouble with being nervous, and if you're nervous about performing magic, trust me, you are not alone. Even as a pro with thousands of hours of performing behind me, I still get nervous once in a while. When I get nervous, my experience allows me to temper my nervousness and get through a potentially challenging performance.

The only thing that truly reduces nervousness is experience and confidence. But if you're terrified about performing, the only way that you can conquer this is to get out there and start performing.

I began my career as a professional magician rather late and I was so nervous at my early performances that my hands would shake. And with unsteady hands and little performing experience, my tendency to make mistakes increased exponentially. But with hindsight, I do have some suggestions that I hope can make your path easier.

The best way to go into a performing situation is to be sure that your magic is thoroughly practiced and nearly second nature. No matter how much I practice, I always find that when I perform a new routine, it always feels rough and unsteady the first time I'm performing it for a crowd. And it's not uncommon for me to make a dumb mistake. However, once I make that mistake, I think about why I made it, make necessary changes, and try to ensure that it won't happen in the future. The stronger your performance and presentation, the better you'll do in front of a crowd.

Try not to perform tricks that you're unsure about or that have a tricky move that makes you uneasy. To gain personal confidence, it's best to perform easy tricks that allow you to focus on presentation. Don't be ashamed if you feel a trick is too easy. If you present a trick well and make it fun, there's no such thing as a bad trick.

Early on, I made the mistake of trying to perform the hardest tricks that I could do. Because I wanted to conquer this nervousness thing, I thought that I could go out there and perform the difficult stuff, gain experience, and build my confidence. However, what happened was that I often made mistakes because I was nervous, which became something of a vicious circle.

If you start out performing easier tricks and work on entertaining audiences, you'll find that your spectators will have fun and be mystified, and you'll become more comfortable standing in front of crowds. And when you're comfortable standing in front of the crowd, you can begin to bring in those harder tricks.

Selecting Tricks and Developing an Act

I THINK THAT MOST MAGICIANS tend to choose and perform tricks that had a large impact on them when they saw it. I know that this is the case with me as I have some tricks in my act that intrigued me when I first saw them on television as a kid.

You can do well performing the tricks that caught your eye, but you also have to think about what's best for your audience. List the tricks that you are learning and get an idea of how they go together. I think it's best to have a variety of material with different types of effects and different interactions with the audience and themes.

To start, you'll want a fairly fast magic trick for your opener to catch the attention of your audience. You want to walk out, command the stage, and perform something fairly quickly. This is not the time for an effect where you write something on five different cards or select seven audience members and bring them onto the stage. You'll have plenty of time in the middle of your act for such setup and interaction.

You'll also want to leave a strong trick, perhaps your best one, for the end of your act. This way, you leave the strongest impact with your audiences. You want your audience to be wowed and excited as they walk away and talk about the magic that they have just witnessed.

As I mentioned earlier, I think it's a good idea to vary your routines. By mixing it up, you maintain your audience's interest. You might have eight great card tricks, but if they involve a spectator selecting a card and you finding it in a similar manner, you're probably not going to make a strong impression.

If you want to see what professional entertainers do, all you have to do is turn on any variety show to see how the acts are mixed up. If there are several singers on the bill, they are not presented one after the other. They are separated by acts that are different. It's much the same with magic.

To start, try to have a minimum of four tricks that you can perform in a row. You may not always perform these tricks together, but you can mix and match them to suit the situation at hand.

The Secrets of the Pros

WHILE PROFESSIONAL MAGICIANS have to practice to maintain their technical skills and when they are developing new routines, the day in and day out rigors of regular performances act to maintain our skills and prepare us for anything audiences may toss out.

Every pro has had a magic trick go bad. And afterwards, most have figured out what went wrong and then tried to remedy the situation so it's less likely to happen again. And every performer has dealt with tough and uncooperative audiences. When less than ideal things happen, it's important to fine tune your next performance to help ensure that the problem doesn't arise again.

Many times, a performer may say something that wasn't scripted that gets a laugh. And in many cases, these quips become a part of the performer's standard routine. Many times, there's almost no way that a performer could have written some of the ad libs that eventually become part of a routine. I know that this has been the case for me.

It's the experience that truly burnishes a routine. The more you perform a routine, the better your performance.

Tricks Gone Bad

SO HOW CAN ONE DEAL with magic catastrophes? Here are some suggestions.

The Trick Blows Up

Do everything you can to salvage a trick, even if it means arriving at an incomplete revelation or ending. Remember that your audience doesn't know what you are going to do so when something goes wrong, you can end it with an intermediate conclusion. Do your best to cover up the mistake and finish, and then move on to your next trick.

If a trick has gone so bad that there's no way to salvage it, don't sweat it. When this happened to me, I simply put the prop away and said something like, "sometimes, things don't go the way that you wish, but..." and I go right into my next trick. This is the reason why it's so important to be able to build an act. Should something go wrong, you can move onto your next trick, a mark of a pro.

An Audience Member Hassles You and Blurts Out a Secret

Many times, when an audience member yells out a secret, they are wrong. I can usually keep going and a later section of the routine will prove them wrong, which will usually quiet the individual. Again, if your presentation is tight and polished, there will be fewer moments for a spectator to consider the secret, and this is less likely to happen.

If an audience member is yelling out the secret, oftentimes other spectators will become irritated with that individual and ask him to stop.

©istockphoto.com/Amanda Rohde

Sometimes, a magician has to consider his or her performing style. If you are performing magic in a confrontational or arrogant manner, you'll be inviting people to be rude to you. But if you go into performances with every intention of entertaining your audiences, most will realize this and be more receptive to your show.

Be gracious to your audiences and they will generally be polite to you. Insult them and be prepared for their wrath.

If an audience member has yelled out the secret and you just can't go on, simply reach a secondary conclusion to your routine, put away the prop, and go to your next trick. It's rare that a lay audience member will be able to explain half of what you can do.

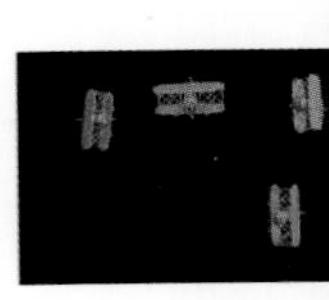

The Rules of Magic

There are cardinal rules in magic; they are as follows:

1. **Never Reveal a Secret to Spectators.**

   When you tell your spectators the secret, you ruin the mystery, which is the basis of your performance. You also ruin the trick for any other magician who may want to perform it.

2. **Never Repeat a Trick.**

   Spectators will often ask you to perform a trick again. Sometimes, it's because they want to enjoy the experience again. But in most cases, it's because they want to watch and have another chance to figure out the secret.

   No matter how many times a spectator asks you to perform a trick again, do not accept the challenge, because that's what it is. The best thing to do in this case is to move onto another trick. And you can do this in a pleasant way by saying something like, "if you liked the last one, here's something else you may enjoy."

3. **Never Tell an Audience What You Are Going to Do.**

   Magic relies on the element of surprise and if you tell an audience what you are going to do, they will be watching you like a hawk to see how you get to the predetermined conclusion. They will be less interested in your performance and entertainment and instead focus on your movements and steps.

   Also, if an audience already knows what you are going to do, there's no surprise when it happens. Many tricks with the greatest impact are those where something happens that the audience is not expecting. You wouldn't give away the ending to a story that you are telling in the beginning. In the same way, you don't want to give away the ending to a trick.

©istockphoto.com/Simon McConico

Learning Magic—DVDs, Books, and Tricks

THE AMOUNT OF INSTRUCTIONAL magic material out there is staggering, and when it comes to learning magic, DVDs and books have their advantages and disadvantages. Before you purchase a DVD, book, or trick, be sure to read reviews if you can before you spend any money.

Evaluating Magic to Buy

A review can tell you if a particular trick will work for you. Some considerations are:

Your Skill Level—do you have the technical skills to perform a particular trick?

The Situations—Is the trick one that you can perform only under certain conditions, such as spectators are in front of you or with certain lighting? Or can a trick be performed while surrounded by a crowd?

Will the Method Fool Spectators?—Believe it or not, there are tricks on the market that are poorly conceived and won't fool anyone. These tricks were developed by someone who simply wanted to make a buck and with a creative mind to create compelling ads.

Close-up or Stand-up?—You have to decide if a trick will work for the types of audiences that you perform for and where you perform. Some tricks require distance to work while others only work in close-up because of the small objects that are used in the effects. Some tricks may require tables or even an assistant to operate something.

©istockphoto.com/Daniel Lemay

You have to find out everything you can before you purchase magic props and tricks. Because the secret to a trick is part of what you're paying for, the manufacturers will not want to tell you the secret. A good reviewer can't reveal a secret, but he can at least let you know under what conditions you can perform a trick.

Every magic enthusiast has a junk drawer that's filled with tricks and instructional materials that he or she never uses. While it's true that it takes a certain amount of exploring and testing to find the tricks that work best for your persona, you want to try and not waste money on magic stuff that you won't use. Buying magic on a whim can be an expensive endeavor.

DVDs

DVDs allow you to see a move or routine in action. And depending on how good the magic teacher is, you should be able to learn the necessary moves.

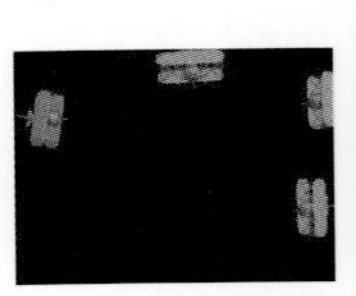

However, the teaching abilities of magicians vary and some offer flat-out inadequate instruction by going too fast or neglecting to discuss important points. And many times, you have to transpose the action that you see on screen when learning a move because the instructor is facing you and his right is your left.

Books generally offer better value. You typically learn more tricks in a book than you can on a DVD. The advantage of a book is that you can open a book and read the explanation at your own pace and a good book features lots of detailed pictures to help you learn. When you want to learn a move, you can set up your hands and then refer to the next page when you're ready.

The disadvantage of a book is that you can't see a move in action as you can on video. It's not unusual to read about a trick or move and then read the explanation and conclude that there is no way that the trick could work. But seeing a move work in video may convince you of its power. Again, try to read reviews before you purchase.

You can read lots of reviews on the internet on sites such as Magic.About.com and Online Visions, and in magic magazines such as *Genii*, *Magic Magazine*, *The Linking Ring* (the official publication of the International Brotherhood of Magicians), and *M.U.M.* (the official publication of the Society of American Magicians).

Buying Magic Materials

There are all kinds of sources for purchasing magic. We'll discuss these as well as their advantages and disadvantages.

The Local Magic Store

If you are lucky enough to have a local magic store in your area, be sure to support it. There is no better way to view magic that you want to buy and obtain advice. A good dealer will advise you not to buy a trick that is too difficult for your skill level and can offer suggestions in its place.

I like going to magic stores because if I'm thinking of purchasing a trick, I can see it demonstrated in the store to access its impact. As you can imagine, there's only so much that you can learn from an advertisement.

While a magic store often charges more than an internet store, the ability to see a magic trick live is worth the extra money that you will spend. A local magic store typically can't offer the broad selection of magic tricks and DVDs that an internet store can, so the selection is often limited. But a magic store can always order products for you.

With the advent of the internet, which offers lots of options for purchasing magic, brick and mortar magic stores have been regularly closing and with the exception of those in tourist destinations, are becoming something of a rarity. It's a valuable resource that is being lost.

The Internet

Internet magic dealers often sell products at the lowest price, but of course, you have to pay for shipping. After shipping, sometimes the price isn't all that different from what you pay at a magic store.

Internet magic stores offer a wide range of tricks and props for sale. But unlike going to a local magic store, you have no way to see a prop in action. Many internet dealers offer videos that you can watch. However, these videos are often shot so the effect looks its best from a certain camera angle and under specific lighting. You can be disappointed by a product when it finally arrives because it won't work for your performing situations.

Magic Clubs

When you become serious about magic, it's important that you meet others with similar interests. While there are local clubs that are sometimes associated with magic stores, the main magic organizations are the International Brotherhood of Magicians and Society of American Magicians, and in England, The Magic Circle. Also, in Los Angeles, the Magic Castle in Hollywood is home to the Academy of Magical Arts.

The International Brotherhood of Magicians (I.B.M.) has local clubs called "Rings" that meet on a regular basis, usually monthly, and feature various activities and lectures. The Society of American Magicians has local clubs called "Assemblies." To find out if there's a club in your local area, you can log onto the websites for these organizations (I.B.M.—www.magician.org and SAM—www.magicsam.com). Both have local clubs in cities and towns across the United States and around the world and hold yearly conventions.

In Conclusion

I hope that you have found this book helpful and established some strong magic fundamentals. Furthermore, I hope that you have been able to entertain your family and friends with magic as a result of the advice and lessons in this book.

Magic is an amazing art that allows you to connect with a wide variety of people. And at its best, you can make people laugh, have fun, and even allow them to experience a sense of wonder.

I wish you the best in your magic endeavors and hope that magic is as fulfilling and fun as it has been for me.

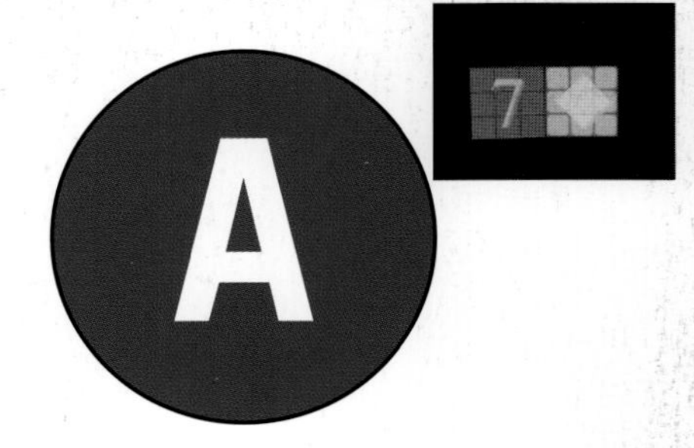

The World's Best
Card Trick

As explained in Chapter 5, this trick involves asking a spectator to read the script out loud in the following section. You follow the directions and make mistakes, but you ultimately reveal the spectator's card. The force card is the two of spades and is displayed on the final page of this appendix.

Place the two of spades on top of your deck of cards and you're ready to go. When the script asks a volunteer to cut the deck, you perform the cross-cut force that's explained in Chapter 4.

Another note, the humor comes in the "mistakes" that occur in the text. Be sure to play these up when you "forget" a prop or a move. For example, one sentence states that you have secretly used a telescope to view the spectator's selected card. Since you don't have a telescope in your hands, you'll get a laugh by looking dumbfounded at your empty hands and it will appear that you have made a huge mistake performing the trick.

Also, when the reader asks if you know the card and other questions, be sure to say "no" and shake your head. At the end, make it appear that you can't possibly know the card.

Also, you may need to explain to a volunteer that he or she needs to wait for you to perform the stated action. Most volunteers will catch on and wait for you to perform the particular task. But some will just read away. It's up to you to control the situation.

Turn the page and then offer the book to the spectator to read.

Welcome to the world of magic.

You're about to embark on an adventure that will allow you to have fun fooling your family and friends.

To perform this trick, you need a deck of cards.

To start, do something fancy with the deck to impress the crowd.

(next page)

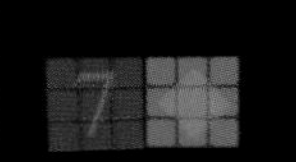

Rest the deck on the table and then ask someone to cut the deck.

(turn page)

Using your superior mind control, you have actually made the spectator cut to a certain card.

(turn page)

Place the card back in the deck and allow the spectator to thoroughly mix the card inside of the rest of the deck.

Have the spectator give the deck back to you.

What the audience doesn't know is that you secretly viewed the card through that telescope that you are carrying in your hand and you now know what the card is.

The card is now in your mind and you're ready to reveal it.

(turn page)

**But before we reveal the card,
we need to build up the suspense.
This is called "entertainment."**

That should be enough.

Now it's time to reveal the card.

Do you know what the card is?

(wait for answer)

(next page)

Sounds like you didn't follow the instructions in this book.

Fortunately, this book comes with a special feature known as the "Magic Backup Function." Using this, you can get out of almost any bad magic situation.

You did pay the fee to subscribe to the "Magic Backup Function," didn't you?

(wait for answer)

Too bad.

Sounds like we're going to have to bail you out anyway for the sake of magic.

Next time, be sure to pay your fees. Do you understand?

(wait for answer)

(turn page and show the audience)

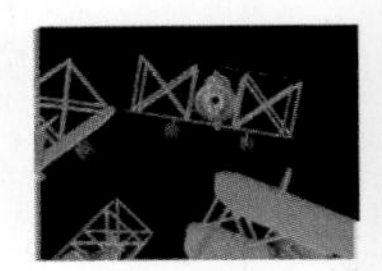

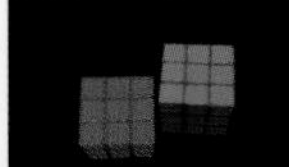

2

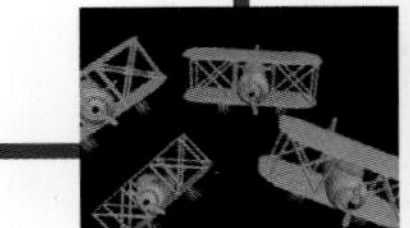

Picture
Yourself as a

Magician 2

THIS TRICK, "PICTURE YOURSELF AS A MAGICIAN 2," is explained in-depth in Chapter 5. To perform the trick, use a cross-cut, Hindu, or cut-deeper force, as taught in Chapter 4, to cause a spectator to select the four of hearts.

Read the lines and then turn the page to show spectators the accompanying gag page. The last page reveals the four of diamonds. For more information, please read about the trick in Chapter 5.

Select a spectator and ask:

Would you believe that your name is written in this book?

(turn page)

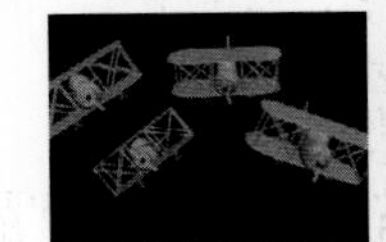

Your
Name

(Name a four-legged animal.)

(Name any country in the world.)

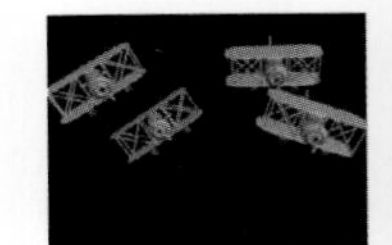

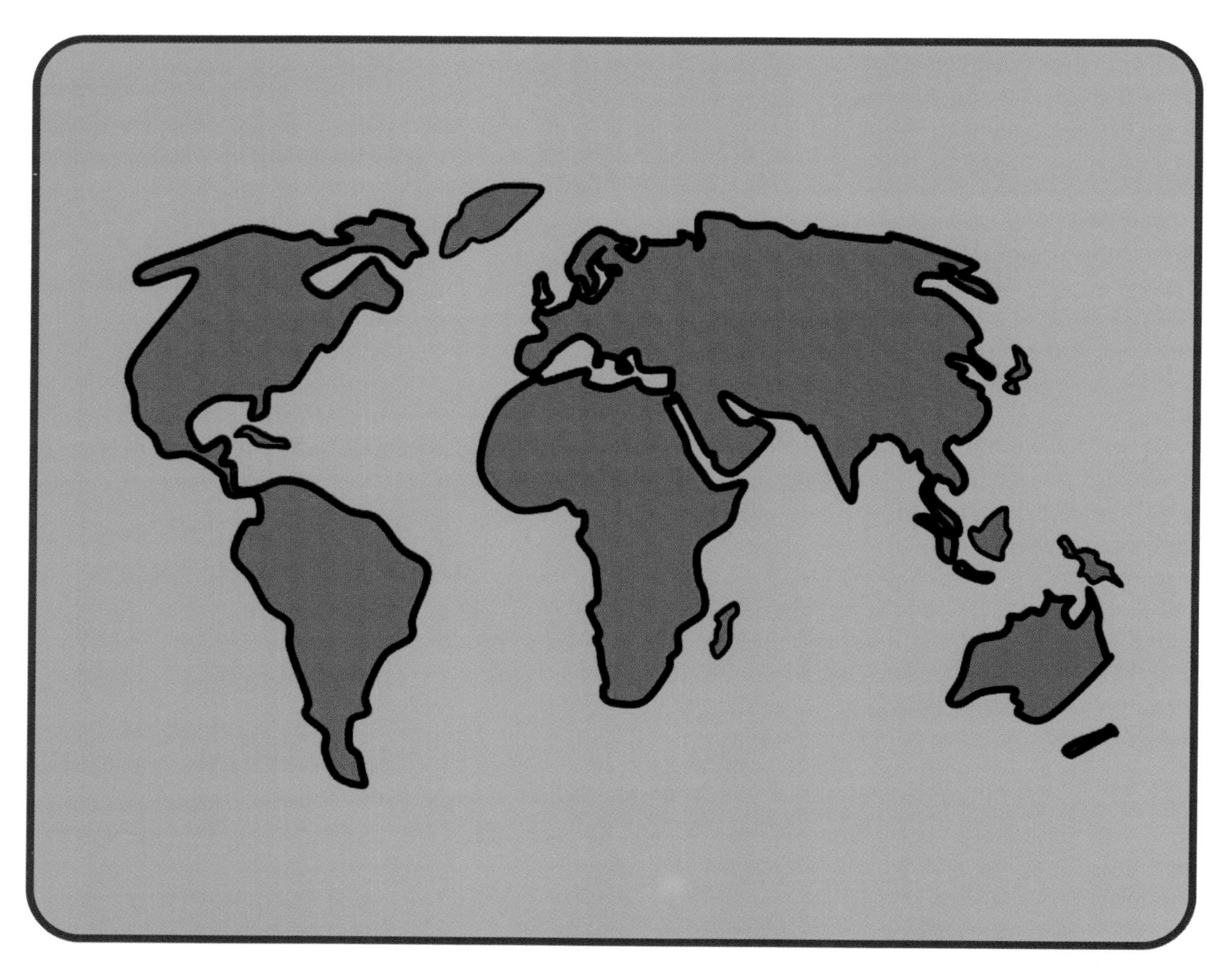

(What was your card?)

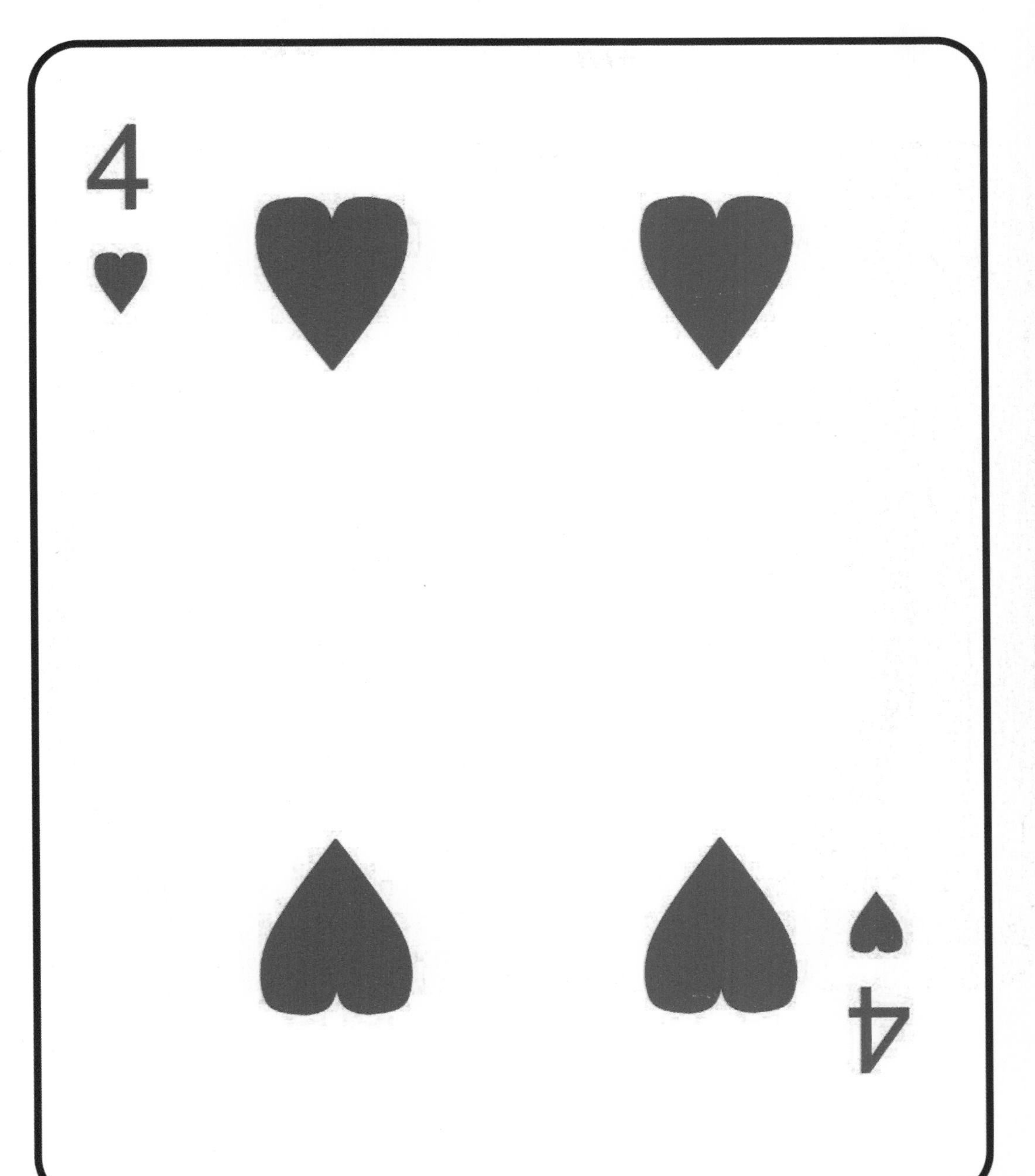

The Clock

Look at the clock below and select any number between 1 and 12.

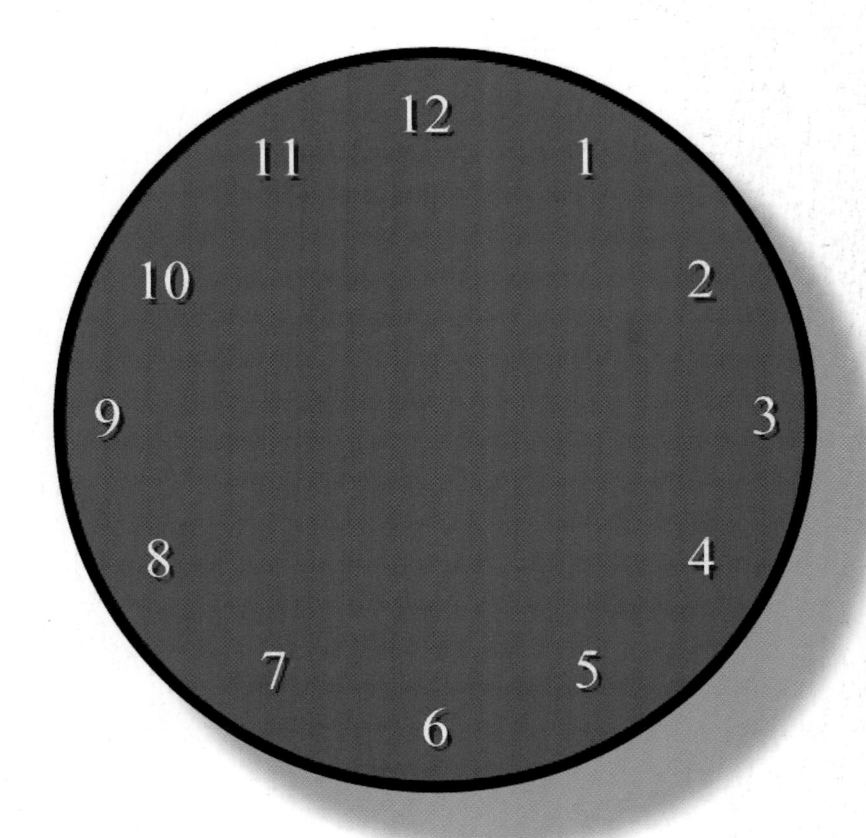

Got a number? Turn the page.

We have drawn lines through the center of the clock.
Find your number and identify the number that's directly across from it, which is joined by the line through the clock's center.

Got a second number?

Subtract the smaller number from the larger number.
Got a new number?

Turn the page.

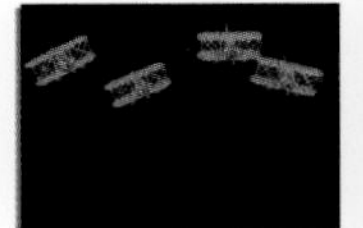

Did you end up with the number 6?

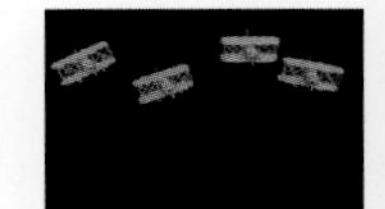

D

The Mind

Reader

Think of a number between two and nine. You can select the number two or the number nine or any number in between.

Multiply your number by nine.

Remember your new number and turn the page.

You should have a two-digit number. Take the individual digits in your number and add them together. For example, if you are now thinking of the number 74, you would add seven and four together to get 11.

Remember your number and turn the page.

Subtract five from your number.

Associate your new number with a letter in the alphabet. For example, if your number is one, associate that with "A," two would be associated with "B," three would be "C," and so on.

Remember your letter and turn the page.

Think of a country that begins with your letter.
For example, if your letter is
"B," you may think of "Brazil."
In the same vein, "D" could be Denmark."
"F" could be France and so on.

Turn the page.

Take the second letter in your country and think of an animal that begins with this letter. Got it?

Turn the page.

Are you thinking of...

An Elephant in Denmark?

Magic

Resources

AS YOUR INTEREST IN MAGIC GROWS, here are resources to help you learn more about magic.

Magic Clubs

THERE'S NO BETTER WAY TO LEARN about magic than to join a club where you can share ideas with your fellow magicians.

Academy of Magical Arts

The Academy of Magical Arts is an exclusive club for magicians and magic enthusiasts that has its headquarters at the Magic Castle in Hollywood, California. The Academy awards the best in magic, bestowing magic's version of the "Oscar" to outstanding magicians each year. The Magic Castle is a nightclub that is a showcase for magicians. You must be referred by a member to visit the Magic Castle or apply for membership to the Academy. Magician membership in the Academy requires an audition.

www.magiccastle.com

The International Brotherhood of Magicians

The International Brotherhood of Magicians (I.B.M.) is a world-wide magic organization that has some 15,000 members. Members meet in local groups called "Rings." There are over 300 Rings in 73 countries. The I.B.M. publishes a monthly magazine, "The Linking Ring."

www.magician.org

The Magic Circle

The Magic Circle is a prestigious magic society that is based in London, England. It has an international membership of some 1500 members.

www.themagiccircle.co.uk

Society of American Magicians

The Society of American Magicians (S.A.M.) has some 250 local clubs called "assemblies" that have been established worldwide. The S.A.M. was formed in 1902 in New York. In 1917, Harry Houdini was elected as its National President, a position that he held until his death in 1926. S.A.M. publishes a monthly magazine, "M-U-M," which represents the organization's motto, "Magic-Unity-Might."

www.magicsam.com

Magazines

MANY MAGAZINES ARE available for magicians.

Genii Magazine

Genii Magazine is a well respected magic publication that is published monthly and features magic advice, articles, tricks, techniques, and reviews.

www.geniimagazine.com

Magic Magazine

Magic Magazine has been published for over 16 years and offers techniques, interviews, articles, and reviews.

www.magicmagazine.com

Magic.About.com

A web-based magazine that's free and part of the About.com family of online publications. Magic.About.com features articles, magic news, interviews, reviews, and beginner's magic tricks.

www.magic.about.com

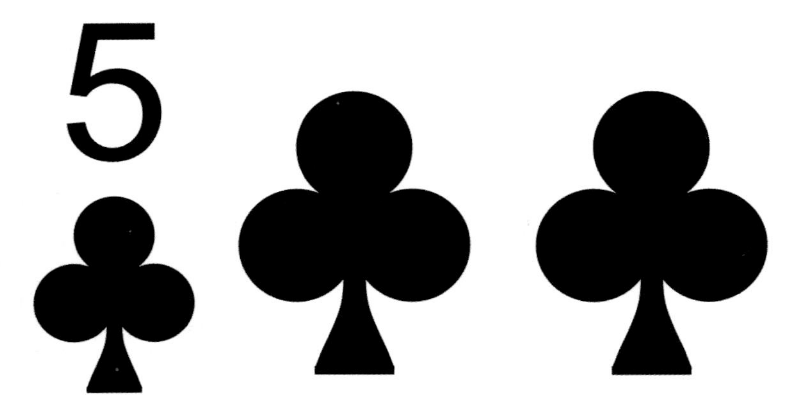
5

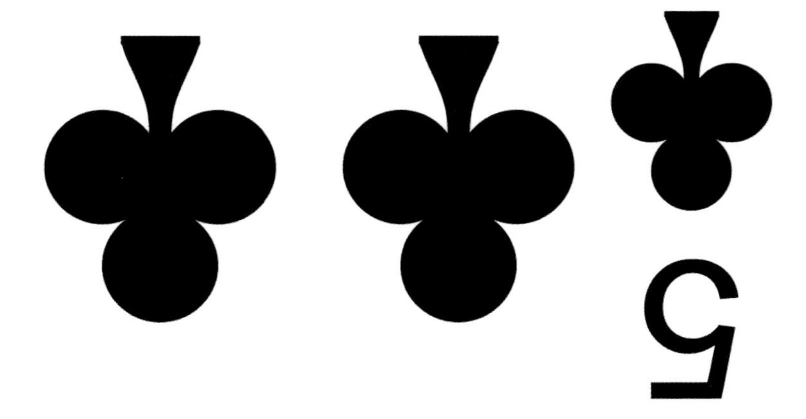
5

Artwork for

"The Missing Pip" from Chapter 9

AS EXPLAINED IN THE INSTRUCTIONS in Chapter 9, the following page offers artwork that you can photocopy, cut out, and then glue to make your giant card. You'll have to make two copies to obtain all of the numbers and pips that you'll need. Please refer to the model in Chapter 9 under "The Missing Pip" for guidance.

As noted in Chapter 9, the artwork on the following page allows you to create a five of clubs. The instructions in Chapter 9 explain how to make a five of diamonds. The five of clubs were chosen for the artwork because it's easier to photocopy black and white artwork.

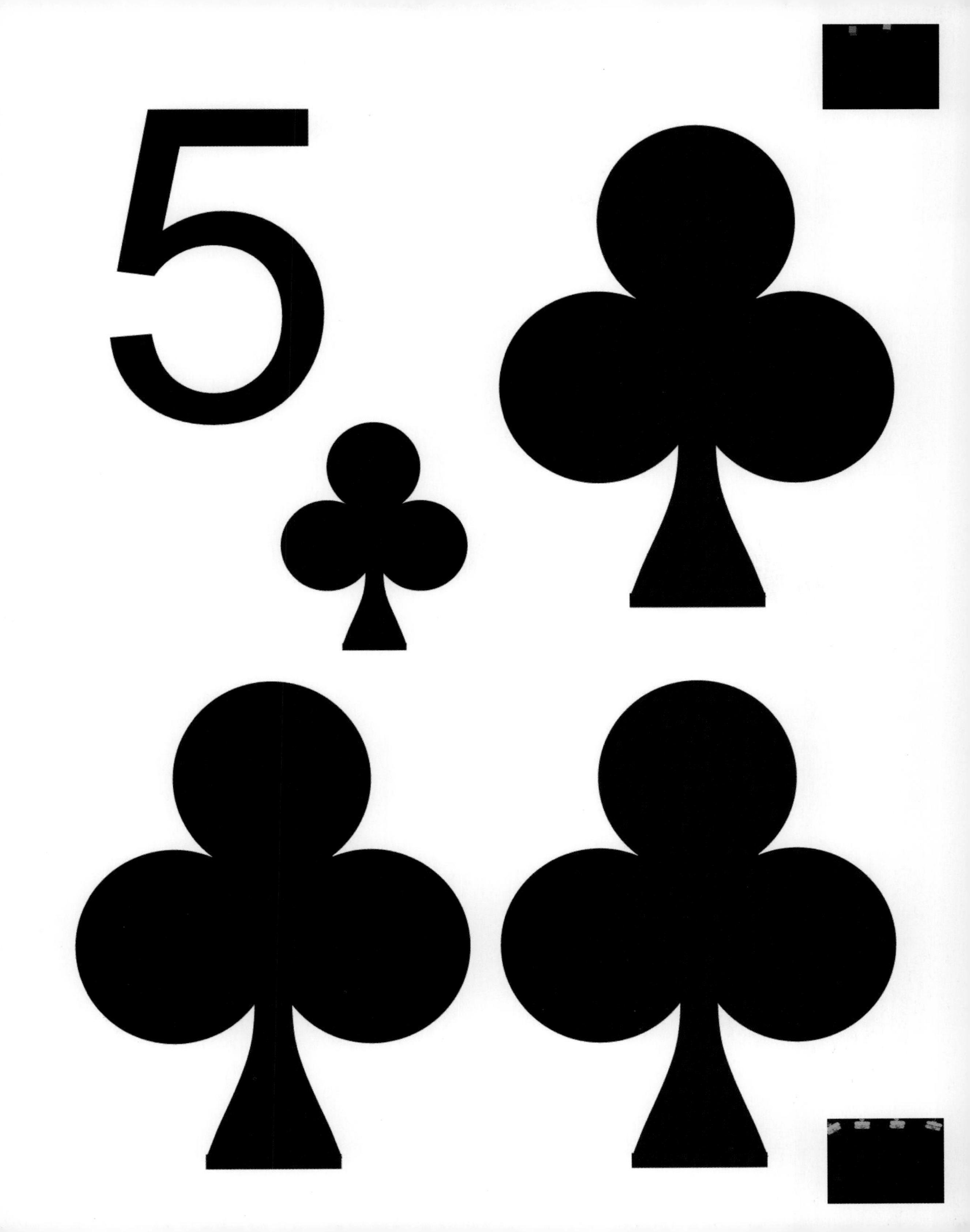

Index

D

E

F

G

H

I

J

K

L

M

T